Sustainable Agriculture and Food Security in an Era of Oil Scarcity

Sustainable Agriculture and Food Security in an Era of Oil Scarcity

Lessons from Cuba

Julia Wright

publishing for a sustainable future

London • New York

First published by Earthscan in the UK and USA in 2009

ISBN: 978-1-84407-572-0 (hbk)
ISBN: 978-0-415-50734-9 (pbk)

Typeset by 4word Ltd, Bristol
Cover design by Yvonne Booth

Cover photo: Cuban landscape, Pinar del Rio © P. Kafno

For a full list of publications please contact:

Earthscan
2 Park Square, Milton Park, Abingdon, Oxon OX14 4RN
Simultaneously published in the USA and Canada by Earthscan
711 Third Avenue, New York, NY 10017
Earthscan is an imprint of the Taylor & Francis Group, an informa business

First issued in paperback 2011

Earthscan publishes in association with the International Institute for Environment and
Development

A catalogue record for this book is available from the British Library

Library of Congress Cataloging-in-Publication Data
Wright, Julia.
  Sustainable agriculture and food security in an era of oil scarcity:
  lessons from Cuba / Julia Wright.
      p. cm.
  Includes bibliographical references and index.
  ISBN 978-1-84407-572-0 (hardback)
  1. Sustainable agriculture--Cuba. 2. Agriculture and energy--Cuba. 3. Agriculture and
state–Cuba. 4. Organic farming–Cuba. 5. Food supply–Cuba. I. Title.
  S477.C9W75 2008
  338.1'87291--dc22
                          2008033034

Contents

List of Figures, Tables and Boxes

Figures

Tables

Boxes

Preface and Acknowledgements

Concerns in the early 21st Century around dwindling oil reserves and their impact on the food system are nothing new and come as no surprise. Way back in the 1970s, for example, school geography classes taught of non-renewable energies, their lifespan and our dependency on them. It was not a fear of oil depletion that inspired the research upon which this book is based, nor any other fear over the state of our food system be it food shortages, human ill-health or environmental degradation. This research was instigated by a deep curiosity to understand why what seemed a more common-sense and logical approach to food and farming was not being practised. For it seemed logical to want to enrich and regenerate the natural resource base for the creation of vibrant, healthy food systems. It seemed logical to want to work in harmony with ecological processes and to avoid destructive activities that continually proved not to work. Yet this logic was not shared by the mainstream, and thus there must be good reasons why not, mustn't there? Thus the author spent over ten years working in agriculture worldwide, with farmers, government officials, researchers and more, looking and listening out for these good reasons. She found none. Then, in 1993, news came through of a situation that might prove once and for all the feasibility of a supposedly more logical farming approach, news of one country that was making a nationwide shift to organic agriculture: Cuba.

Several years of planning and organising culminated in the execution of doctorate research on the coping strategies of the Cuban people as they attempted to feed themselves after their main suppliers of food and fuel had been permanently cut off with the dissolution of the Socialist Bloc in 1989. During the period of the early to mid 1990s, news from Cuba comprised mainly anecdotal stories from short-stay visitors to the country, visitors who were exposed only to the emerging successes of urban agriculture. Little news was coming through of what was going on in rural areas, or of the food system in general. This book attempts to fill that gap. It also turns out to be a snapshot from an era that may not be repeated, for into the new Millennium has come a slight but steady increase in prosperity, changes in governance, and even the prospect of national oil reserves for Cuba, all of which have served to diminish the country's need to implement alternative farming and food systems.

It came as no surprise to find that Cuba was not, as a whole, farming organically. An absence of fuel and agrochemicals is not synonymous with organic agriculture. The enigma of Cuba has provided great inspiration and vision for communities all over the world intent on developing more localised, sustainable food systems. Yet dispelling the organic myth is important in order to learn from the country's real experiences and so to enable other nations and communities to move further toward their goals. That said, the Cuban experience is interpreted and written through the eyes of the author, a foreigner. During the 1990s, rural social sciences were little developed in Cuba and studies such as this one were not undertaken by Cuban nationals. This situation has since changed, and more nuanced and personal accounts of the period will no doubt ensue.

Acknowledgement is due to everyone who advised, facilitated, contributed and got involved in some way over the lengthy period of producing this work, in Cuba, the Netherlands and England. The three farsighted enablers were Professor Niels Röling of Wageningen University, the Netherlands, Dr Humberto Ríos Labrada of the Instituto Nacíonal de Ciencias Agrícolas, Havana, and the European Commission, which came up with the initial research grant. Other key people helped enormously; they know who they are.

List of Acronyms and Glossary

Acronyms and names of Cuban institutions

ACAO Asociación Cubana de Agricultura Organica
   Cuban Association of Organic Agriculture

Acopio State food collection and distribution agency (National Union of)

ACPA Asociación Cubana de Producción Animal
   Cuban Association of Animal Production

ACTAF Associación Cubana de Tecnicos Agricolas y Forestales
   Cuban Association of Agricultural and Forestry Technicians

ANAP Asociación Nacional de Agricultores Pequeños
   National Association of Small Farmers

BANDEC Banco de Credito y Comercio
   Credit and Commerce Bank

CAI Centro AgroIndustrial
   Agroindustrial Centre for Sugarcane

CATEC Comercial Agropecuaria Export-Import
   Agricultural import-export marketing company

CCS Cooperativa de Creditos y Servicios
   Credit and Service Cooperative

CEAS Centro de Estudios de Agricultura Sostenible
   Centre for Sustainable Agriculture Studies

CIGB Centro de Investigaciones Genetica y Biotechnologia
   Genetics and Biotechnology Research Centre

CIGEA Centro de Informacion, Gestion y Educaciona Ambientales
   Centre for Environmental Information, Education and Management

CIPS Centro de Investigaciones de Psicología y Sociología
   Psychology and Sociology Research Centre

CITMA Ministerio de Ciencia, Tecnologia y del MedioAmbiente
   Ministry of Science, Technology and the Environment

Citricus Caribe State citrus marketing entity

CPA Cooperativa de Producción Agricola
   Agricultural Production Cooperative

CPMOL	Centros Procesadores de Materia Organica y Lombricultura
	Processing Centres for Organic Matter and Vermiculture
CREE	Centro por la Reproducción de Entomofagos y Entomopatogenos
	Centre for the Reproduction of Entomophages and Entomopathogens
DECAP	Consejo de Iglesias de Cuba, Departmento de Coordinación y Asesoria de Proyectos
	Cuban Council of Churches
ETIAH	Estacion Territorial de Investigaciones Agricolas de Holguin
	Holguin Regional Agricultural Research Station
ETPP	Estaciones Territoriales de Protección de Plantas
	Regional Stations of Plant Protection
FNH	Fundación de la Naturaleza y el Hombre 'Antonio N. Jimenez'
	'Antonio N. Jimenez' Foundation for Nature and Humanity
Frutas Selectas	State food collection and distribution agency for tourist markets
ICA	Instituto de Ciencias Animales
	Institute for Animal Sciences
INCA	Instituto Nacional de Ciencias Agricolas
	National Institute of Agricultural Sciences
INHA	Instituto de Nutricion y Higiene Alimentaria
	Institute of Food Hygiene and Nutrition
IIES	Instituto de Investigaciones de Ecologia y Sistematica
	Research Institute of Ecology and Systematics
IIHLD	Instituto de Investigaciones Horticolas 'Liliana Dimitrova'
	'Liliana Dimitrova' Horticultural Research Institute
IIMA	Instituto de Investigaciones de Mecanizacion Agricola
	Research Institute of Agricultural Mechanization
IIPF	Instituto de Investigaciones de Pastos y Forages
	Pastures and Forages Research Institute
IIRD	Instituto de Investigaciones de Riego y Drenaje
	Irrigation and Drainage Research Institute
INIFAT	Instituto Nacional de Investigaciones Fundamentales de Agricultura Tropical
	National Research Institute for Tropical Agriculture
INISAV	Instituto Nacional de Investigaciones de Sanidad Vegetal
	National Research Institute of Plant Health
INIVIT	Instituto Nacional de Investigaciones de Viandas Tropicals
	National Research Institute for Tropical Roots and Tubers
ISCAH	Instituto Superior de Ciencias Agricolas de Havana (renamed – see UNAH)

MINAG	Ministerio de la Agricultura (also known as MINAGRI)
	Ministry of Agriculture
MINCIN	Ministerio de Commercio Interior
	Ministry of Domestic Trade
ONE	Oficina Nacional de Estadisticos
	National Statistics Office
Sanidad Vegetal	National Institute of Plant Protection
SINCITA	Sistema Nacional de Ciencias Agricolas y Innovación Technologica
	National System of Agricultural Science and Technological Innovation
SISVAN	Sistema Nacional de Vigilancia Alimentaria y de Nutricion
	National Food and Nutrition Surveillance System
UBA	Unidades Basicas de Agroalimentos
	Basic Food Supply Unit
UBPC	Unidades Basicas de Produccion Cooperativa
	Basic Units for Cooperative Production
UCLV	Universidad Central de Las Villas
	Central University of Las Villas
UNAH	Universidad Nacional de Agricultura de Habana
	National Agricultural University of Havana
UNH	Universidad Nacional de Habana
	National University of Havana

Non-Cuban acronyms

CGIAR	Consultative Group on International Agricultural Research
CMEA	Council of Mutual Economic Assistance
EU	European Union
FAO	Food and Agriculture Organization (of the UN)
FHIA	The Honduran Agricultural Research Foundation
IFOAM	International Federation of Organic Agriculture Movements
NGO	Non-governmental organization
UNDP	United Nations Development Programme
USAID	United States Agency for Intenational Development
WHO	World Health Organization (of the UN)

Glossary and translations

agro-ecology   the scientific basis of alternative agriculture, synonymous
                in some circles with collective social action on the

	eco-management of the natural resource base; common usage in Latin America
bagasse	fibre remaining after the extraction of the sugar-bearing juice from sugarcane
cachaza	sugarcane waste
campesinos	traditional, peasant, small-scale, autonomous farmers
canasta basica	basic state ration
comercio minorista	retail trade network
consejos populares	People's Councils
industrialized agriculture	farming system based on concepts of industrialization – high external inputs, high outputs, homogeneous environment and people, tending to monocropping and large-scale systems
La Bodega	food ration store
La Libreta	food ration book
Las Placitas	capped-price municipal markets
latifundio	large landed estate
minifundio	smallholding
organic agriculture	farming system that uses regenerative husbandry practices based on the principles and science of ecology; synonymous with agro-ecology, ecological, biological and biodynamic agriculture, and natural farming
organoponico	raised-bed, intensive urban agricultural unit
tiro directo	direct marketing
en usufructo	in perpetuity

Cuba: Providing the Model for a Post-Petroleum Food System?

Petroleum-based food systems and food security

Over the next few decades, nations will be experiencing fluctuations and increasing scarcity of fossil fuel supplies, and this will affect food prices. Alternative farming and food systems are required. Industrialized countries in particular have been over-consuming fossil fuels by two-thirds, and their agricultural sectors have contributed this with their heavy dependence on cheap fossil energy for mechanization and as a basis for agrochemical inputs such as pesticides and fertilizers. The corresponding industrial food systems in which these farming systems are embedded are similarly dependent on cheap fossil fuels for the ever-increasing processing and movement of foodstuffs. The low fuel prices, combined with the industry's avoidance of paying clean-up costs of environmental pollution, have enabled the maintenance of low food prices (Vandermeer et al, 1993; Odum, 1994; Tansey and Worsley, 1995; Desai and Riddlestone, 2002; Harrison, 2004). Alternative, organic agriculture shows to perform better on a per hectare scale with respect to both direct energy consumption (fuel and oil) and indirect consumption (synthetic fertilizers and pesticides) (Scialabba and Hattam, 2002; Ziesemer, 2008). Many of the products of organic farming are processed and marketed through the industrial food system, but their prices are higher owing to their factoring-in of their impacts on the environment (Pretty et al, 2000). Although research has long been under way into energy alternatives, the agriculture and food sectors make little advance in developing alternative systems as long as fuel prices remain low.

A far cry from these petroleum-dependent populations are the 90 per cent of the world's farmers who manage 75 per cent of global agricultural lands and who have little recourse to fossil fuels and inputs (Conway, 1997). For many of these farmers, their low-input and organic status is by default rather than choice. Yet others have opted out of the opportunity to embrace industrialized, Green Revolution agriculture when offered to them.[1] Should these farmers, and the food systems of their countries, be encouraged to take the industrialized route and to also depend on fossil fuels, or might they leap-frog into developing more efficient and effective alternative food systems?

Yet more localized, petroleum-free farming approaches are perceived by many as unable to deliver the yields required to feed growing populations, especially those of agrarian-based countries. An estimated 200 million people are classified as undernourished in Africa alone, and with forecasts predicting a shortfall in meeting the Millennium Development Goal of halving global food insecurity by 2015, pressure remains on the agricultural sector to increase yields (FAO, 1998; IAC, 2003; Benson, 2004). Evidence is mounting that alternative farming approaches can outperform industrialized farming in many circumstances (e.g. Pretty, 1998; Parrott and Marsden, 2002; Scialabba and Hattam, 2002; IFAD, 2003). However, this evidence is piecemeal and small-scale. No single country has made a policy commitment to, and effected, a nationwide sustainable, organic production approach. Thus there is no example of what a post-petroleum food system might look like, nor how to put this into place in terms of research, extension and policy support (Röling and Jiggins, 1998).

Cuba: the global example of a post-petroleum food system?

Or is there? Throughout the 1990s, reports were coming through on the resounding success of Cuba in heading-off a major national food crisis, a crisis that had been brought on by drastic shortfalls in imported fuel, food and agrochemical input supplies (these reports include, for example, Levins, 1990; Altieri, 1993; Carney, 1993; Rosset and Benjamin, 1994; Wilson and Harris, 1996; Rosset and Moore, 1997; Ritchie, 1998; Bourque, 1999; Moskow, 1999; Murphy, 1999). Figure 1.1 lists some of the headlines of these reports which emerged mainly through study tours[2] and visits to Cuba by foreign interest groups. Was Cuba demonstrating a post-petroleum food system, feeding its people through state-supported, localized, organic farming?

The Cuban food system in crisis

Ever since its Socialist Revolution of 1959, Cuba has maintained restricted and selective contact with non-socialist countries. From an international perspective, this has resulted in a relative dearth of knowledge on all aspects of Cuban life, and a heavy reliance on anecdotal evidence. Nevertheless, there was no doubt that the dissolution of the Socialist Bloc of the Soviet Union and Eastern Europe in 1989 had brought an abrupt end to the support it had provided Cuba, and with this went the inputs that Cuba had relied upon to maintain its highly industrialized system of agriculture – petrol, machinery, chemical fertilizers and pesticides.

"CRISIS FORCES CUBA TO CONSERVE ...
In an attempt to ride out the difficult times, the Government is becoming increasingly "green"..."

Kaufman, 1993

"The greening of the revolution: Cuba's experiment with organic agriculture."
Rosset & Benjamin, 1994

"Cuba Goes Organic! ...
Pushed by the loss of imported pesticides and fertilisers and pulled by a growing awareness of environmental damage caused by intensive agricultural techniques; the Cuban government looked to sustainable, organic methods of cultivation to resuscitate and develop domestic food production."

Weaver et al, 1997

"Food Security and Local Production of Biopesticides in Cuba.
Cuba has embarked on an unprecedented national transition from high-external input to low-input and organic agriculture ..."

Rosset & Moore, 1997

"A revolution in Urban Agriculture ...
Cubans have combined the revival of traditional farming systems with the development of high-tech methods."

Ritchie, 1998

"CUBA'S ORGANIC REVOLUTION.
Organic agriculture has been adopted as the official government strategy for all new agriculture in Cuba ..."

The Pesticides Trust, 1998

"Fidel's sustainable farmers."
Economist, 1999

"Cuba: towards a national organic regime?
Today, 65% of Cuba's rice and 50% of its fresh vegetables are produced organically."

Parrott & Marsden, 2002

"An Organic Coup in Cuba?
Castro says organic is working."

Simon, 1997

Figure 1.1 *International reporting on the transformation towards organic agriculture in Cuba in the 1990s*

From the beginning of the Revolution, Cuba had been influenced by the ideology of industrialization, in order to emancipate the rural population from the perceived drudgery of hand labour and to provide an abundant supply of cheap food. In the words of the Communist Manifesto, this was undertaken through 'a triumphant conquering and domination of nature by man'. Cuba and the other socialist countries in the Council of Mutual Economic Assistance (CMEA) had relied on each other for obtaining whatever goods and services they required, this being necessitated by the trade sanctions of politically opposed countries. Cuba had then followed a model of externally assisted agricultural modernization, one with a far more rapid rate and spread than of other agrarian-dependent countries. This industrial growth was dependent on imported inputs and capital goods (Pastor, 1992). Cuba, a country without significant petroleum reserves, was in receipt of subsidized imports of fuel, agrochemical inputs, technology, training, many basic foodstuffs and medicines. In return it sold its tropical produce – largely sugarcane but also tobacco and fruits (along with other resources such as nickel) – to those temperate-climate socialist countries at more than double the world market price. This symbiotic trading relationship lasted for three decades, from the 1960s to the 1980s,[3] and endured so well that the materially abundant decade of the 1980s was subsequently referred to in Cuba as 'the years of the fat cow'.[4]

The abrupt changes of 1989 hit the Cuban agricultural sector particularly hard for four reasons. Firstly, Cuba had an extreme industrialized agricultural system, one that was using more tractors and applying more nitrogen fertilizer per hectare (192kg/ha) than similar production systems in the USA (Hamilton, 2003). Mechanized irrigation systems covered over one-quarter of all crop land. Secondly, Cuba was importing not just a select few of the inputs and foodstuffs it required for survival but the large majority of them. In 1988, for example, 90 per cent of fertilizers and pesticides, and 57 per cent of food needs, were being imported (Rosset and Benjamin, 1993). Within the country, farms controlled by the Cuban Ministry of Agriculture, which worked 70 per cent of agricultural lands, were producing just 28 per cent of nationally consumed calories. Thirdly, not only did Cuba lose its Soviet trading partners, who were paying preferential prices for its products – an average of 5.4 times the world market price for sugar, for example – but just as Cuba was forced to enter the global sugar market, international commodity prices plummeted.[5] Finally, over the previous 30 years Cuba had developed very little in the way of its own diversified agricultural products or light industry, either for export or for domestic consumption (Pastor, 1992).

The years of the fat cow were over. Between 1990 and 1993, according to reports, the availability of pesticides and fertilizers fell around 80 per cent, while fossil fuel supply dropped by 47 per cent for diesel and 75 per cent for petrol. Electricity levels also then fell, by 45 per cent. Agricultural production and food availability dropped to critical levels, with average calorific intake falling by as much as 30 per cent compared to previously and food imports dropping by over 50 per cent (PNAN, 1994; Rosset and Moore, 1997; Rosset, 2000).

The reported response to the crisis

In 1990, the state declared the start of a 'Special Period in Peace Time', a self-imposed state of emergency which urged the need for sacrifices in living standards, including an acceptance of insufficient food supplies, in order to buy the country time to build up its levels of self-sufficiency and particularly to meet basic food requirements (Rosset and Moore, 1997).[6] Within this framework, the agricultural sector was tasked with finding solutions to production problems, and to do so using local resources. Given the importance that the Revolution placed on science and technology, this food security mandate was spearheaded by the scientific community.[7] Researchers who had previously been beavering away in isolation on the development of alternative technologies were now mobilized and brought into the mainstream, and already-existing plans to produce organic pesticide and fertilizer products were put into operation and scaled-up in order to replace the shortfall of imported chemical inputs. In place of tractors, traditional teams of oxen were reinstated, and the knowledge and skills of older farmers were sought for the handling of the livestock as well as for other issues.

One major chronicler of this period, Rosset (2000, p206), had the following to say about the change in agricultural technologies:

> In response to the crisis, the Cuban government launched a national effort to convert the nation's agricultural sector from high input agriculture to low input, self-reliant farming practices on an unprecedented scale. Because of the drastically reduced availability of chemical inputs, the state hurried to replace them with locally produced, and in most cases biological, substitutes. This has meant bio-pesticides (microbial products) and natural enemies to combat insect pests, resistant plant varieties, crop rotations and microbial antagonists to combat plant pathogens, and better rotations and cover cropping to suppress weeds. Synthetic fertilizers have been replaced by bio-fertilizers, earthworms, compost, other organic fertilizers, natural rock phosphate, animal and green manures, and the integration of grazing animals. In place of tractors, for which fuel, tyres and spare parts were largely unavailable, there has been a sweeping return to animal traction.

The area of change that gained the most international coverage and interest was that of urban agriculture (Weaver, 1997; Murphy, 1999). As the disastrous impacts of the import shortages grew more visible, the state decreed that all fallow and unused urban land be cultivated in perpetuity (*en usufructo*) and free from taxes. People from all professions took up this opportunity and, supported by the state, developed an intensive network of cultivated plots. By 1998,

Havana had more than 26,000 urban gardens, producing 540,000 tons of fresh fruits and vegetables (Moskow, 1999).

For agriculture as a whole, yields of many basic food items increased, in some crops to levels higher than those of the previous decade, especially those of roots, tubers and fresh vegetables (Rosset, 1998; Funes, 2002). The food crisis had, according to reports, been lessened or even overcome. As Rosset (1996, p66) explains: 'Although no figures are available, numerous interviews and personal observations indicate that by mid-1995 the vast majority of Cubans no longer faced drastic reductions of their basic food supply.'

It was not only agricultural production that had apparently been transformed. According to reports, the Cuban government had succeeded in maintaining its socialist policy of feeding its people. Cuba had historically placed high priority on social concerns and had invested in the development and provision of education, communication channels, housing and health care facilities. It was this solid foundation that provided the bedrock for Cuba's survival (Rosset and Moore, 1997). Just over mid-way through the decade, Fidel Castro announced (1996): 'We can proudly say that despite the difficult circumstances, we were able to ensure equal access opportunities for the entire population to the available food, health and education.'

A snapshot of the Cuban experience in the 1990s

The story of the way in which Cuba developed a post-petroleum food system has endured into the 21st century (see for example, Oppenheim, 2001; Hamilton, 2003; Snyder, 2003; Handscombe and Handscombe, 2004). In fact, in the latter half of the 2000s, Cuba was still heralded as the chief example of success in the debate around peak oil and the development of more sustainable alternatives (see, for example, Pretty, 2005; Hopkins, 2006; Pfeiffer, 2006; Heinberg 2007; Sligh and Christman, 2007). Yet there has been little in the way of analytical or evaluative study on the experiences of Cuba by researchers and policy-makers concerned with the depletion of fossil fuel reserves and/or sustainable food-security solutions,[8] including by those who denounce localized organic approaches as being inappropriate on a large scale. Within Cuba, evaluations and data were also scarce. In one attempt to assess the economic and social performance of the Cuban transition in the 1990s, Mesa-Lago (1998) notes in relation to Cuban data, 'Statistical series vanished at the beginning of the transition, making a serious evaluation virtually impossible. In 1995–97, important data were released but their reliability is questionable.'[9] In addition, socio-economic research in Cuba was considered a matter of national interest and results were frequently unavailable for wider international dissemination. Asssessing the performance of the country in coping with depleted oil resources was not considered a national research priority over other pressing issues of the period. Given this absence of information from Cuban organizations – or of any systematic, broad-scale analysis during the mid- to late-1990s – it is the largely anecdotal reports that form the basis for the perception that Cuban agriculture was feeding its people through a self-sufficient, organic style of agriculture.

Nevertheless, whether or not the reports were true, the experiences of Cuba in coping with the rapid demise of the Socialist Bloc and thereby its source of food and fuel supplies ought to provide valuable lessons for other nations vulnerable to similar predicaments, whether occurring virtually overnight, as happened in Cuba, or as part of a longer-term decline.

This book provides a snapshot of the Cuban food and farming system as it struggled to cope with continued low-levels of petroleum inputs and food imports during the mid-to-latter stages of the 1990s. It is based on research carried out in Cuba during 1999 to 2001. After that period, petroleum and food imports have been steadily increasing into the country. The experiences of the 1990s thus stand unique, for the time being at least.

The book first addresses the concept of peak oil and the dependency of the industrialized food and farming system, typified by Cuba. As well as developing and implementing alternative energy supplies, the peak oil challenge provides the opportunity for more creative improvements of the farming and food system in order to achieve equitable access and optimal health for all and over the long term. An analysis of recent Cuban agricultural history explains why the country found itself in such a predicament of dependency at the end of the 1980s. The experiences of Cuba in coping with the lack of petroleum inputs is assessed from three perspectives: the resultant food system emerging in Cuba, the coping strategies that were required to ensure food security, and the degree to which Cuban agriculture is run along organic lines. Based on these perspectives, lessons are drawn out on the applicability of the Cuban experience for the development of more secure, localized, organic and equitable food and farming systems worldwide.

Notes

1   This group includes the 2.5 million farmers with over 5 million ha who intentionally follow organic principles (Altieri et al, 2000).

2   The author herself participated in two such tours, in 1995 and 1999.

3   In fact, Cuba's external dependency on agricultural inputs, food imports and sales of its main crop – sugar cane – originated long before the Revolution of 1959, such dependency being a fairly standard feature of trade relations between the USA, Europe and much of the Caribbean region (and the developing world) since the beginning of the 20th century (Enríquez, 2000).

4   During this time, farmers were paid on the basis of how much they produced, no mat-
    ter what the production cost, and workers on the basis of number of days worked, no
    matter what their output (Sinclair and Thompson, 2001).
5   This was exacerbated by the trade sanctions concurrently imposed on Cuba by the
    United States.
6   In 2008 this Special Period was still in place.
7   A much quoted figure from Rosset and Benjamin (1993) is that Cuba has only 2 per
    cent of the population of Latin America, but 11 per cent of its scientists.
8   Notwithstanding the handful of invaluable, individual efforts, such as Deere, 1997;
    Lane, 1997; Mesa-Lago, 1998; Enriquez, 2000; and the work of Rosset at the Institute
    for Food Policy Research, for example Rosset and Benjamin, 1994; Rosset, 1996.
9   More detailed reasons for concern over the validity of Cuban statistics is provided by
    Alvarez, 1994, pp60–66.

References

Altieri, M. A. (1993) 'The implications of Cuba's agricultural conversion for the general
    Latin American Agroecological Movement', *Agriculture and Human Values*, vol 10, no
    3 (Summer 1993), pp91–92
Altieri, M. A., Rosset, P. and Thrupp, L. A. (2000) *The Potential of Agroecology to Combat
    Hunger in the Developing World*. www.cnr.berkeley.edu/~agroeco3/ accessed in
    August 2008
Alvarez, J. (1994) *Cuba's Infrastructure Profile*, International Working Paper Series IW94-
    4, July 1994. Food and Resources Economics Department, Institute of Food and
    Agricultural Sciences, University of Florida, Gainesville
Benson, T. (2004) *Africa's Food and Nutrition Security Situation. Where Are We and How
    Did We Get Here?* 2020 Discussion Paper No. 37, IFPRI. www.ifpri.org/2020/
    dp/dp37.htm accessed in December 2004
Bourque, M. (1999) 'Policy options for urban agriculture', *Conference Proceedings,
    Growing Cities Growing Food: Urban Agriculture on the Policy Agenda*, October 1999,
    Havana, Cuba. Institute for Food and Development Policy (Food First)
Carney, J. (ed) (1993) 'Low-Input Sustainable Agriculture in Cuba', *Agriculture and
    Human Values*, vol 10, no 3 (Summer 1993)
Castro, F. (1996) Speech at UN World Food Summit, 18 November 1996
Conway, G. (1997) *The Doubly Green Revolution, Food for all in the 21st Century*, London,
    Penguin Books
Deere, C. D. (1997) 'Reforming Cuban agriculture', *Development and Change*, vol 28,
    pp649–669
Desai, P. and Riddlestone, S. (2002) *Bioregional Solutions for Living on One Planet*,
    Schumacher Briefing No. 8, Green Books
Economist (1999) 'Fidel's Sustainable Farmers', *The Economist* 1351(8116) (April 24–30),
    p34
Enríquez, L. J. (2000) *Cuba's New Agricultural Revolution. The Transformation of Food
    Crop Production in Contemporary Cuba*. Development Report No. 14, Dept.
    Sociology, University of California. http://www.foodfirst.org/node/271
FAO (1998) 'Evaluating the potential of organic agriculture to sustainability goals', FAO's

Technical Contribution to The *IFOAM Scientific Conference*, Mar del Plata, Argentina, November 1998

Funes, F. (2002) 'The organic farming movement in Cuba', in Funes, F., Garcia, L., Bourque, M., Pérez, N. and Rosset, P. (eds) *Sustainable Agriculture and Resistance: Transforming Food Production in Cuba*, California, Food First Books, pp1–26

Hamilton, H. (2003) *A Different Kind of Green Revolution in Cuba*, Oakland CA, Food First/Institute for Food and Development Policy. http://www.sustainer.org/pubs/columns/03.24.03Hamilton.html accessed September 2004

Handscombe, C. and Handscombe, D. (2004) 'Cuba saved by smallholders', *Smallholder*, January, pp66–67

Harrison, A. (2004) 'Over the Fence', *Landmark* Issue 54 (January/February), p3

Heinberg, R. (2007) *The Party's Over: Oil, War and the Fate of Industrial Societies,* London, Clairview Books

Hopkins, R. (2006) *Energy Descent Pathways: Evaluating Potential Responses to Peak Oil*, MSc thesis, University of Plymouth. www.transitionculture.org

IAC (2003) *IAC Study on Science and Technology Strategies for Improving Agricultural Productivity and Food security in Africa*. Progress Report, May 2003 (unpublished), Interacademy Council Study Panel

IFAD (2003) *The Adoption of Organic Agriculture Amongst Small Farmers in Latin America and the Caribbean: Thematic Evaluation*. IFAD Report No. 1337, Rome, IFAD

Kaufman, H. (1993) 'From red to green: Cuba forced to conserve due to economic crisis', *Agriculture and Human Values*, vol 10, no 3 (Summer 1993), pp31–32

Lane, P. (1997) 'El modelo Cubano de desarrollo sostenible', *Seminario Internacional Medio Ambiente y Sociedad*, Havana

Levins, R. (1990) 'The struggle for ecological agriculture in Cuba', *Capitalism, Nature and Socialism*, vol 5, pp121–141

Mesa-Lago, C. (1998) 'Assessing economic and social performance in the Cuban transition of the 1990s', *World Development*, vol 26, no 5, pp857–876

Moskow, A. (1999) 'Havana's self-provision gardens', *Environment and Urbanisation*, vol 11, no 2, pp127–132

Murphy, C. (1999) *Cultivating Havana: Urban Gardens and Food Security in the Cuban Special Period*. Unpublished Masters thesis, Fac. Latinoamericana de Ciencias Sociales, Universidad Nacional de Habana.

Odum, H. T. (1994) 'Energy analysis of the environmental role in agriculture', in Stanhill, G. (ed) *Energy and Agriculture*, Berlin, Springer Verlag, pp24–51

Oppenheim, S. (2001) 'Alternative agriculture in Cuba', *American Entomologist*, vol 47, no 4 (Winter 2001), pp216–227

Parrott, N. and Marsden, T. (2002) *The Real Green Revolution. Organic and Agroecological Farming in the South*, London, Greenpeace Environmental Trust

Pastor, M. (1992) *External Shocks and Adjustment in Contemporary Cuba*. Working Paper, International & Public Affairs Centre, Occidental College, USA

Pesticides Trust (1998) 'Cuba's organic revolution', Eco-Notes. www.ru.org/81econot/html accessed May 2000

Pfeiffer, D. A. (2006) *Eating Fossil Fuels. Oil, Food and the Coming Crisis in Agriculture*, Canada, New Society Publishers

PNAN (1994) *Plan Nacional de Acción Para la Nutrición*, Havana

Pretty, J. (1998) *The Living Land*, London, Earthscan Publications

Pretty, J. (2005) *The Pesticide Detox: Towards a More Sustainable Agriculture*, London, Earthscan Publications

Pretty, J., Brett, C., Gee, D., Hine, R. E., Mason, C. F., Morison, J. I. L., Raven, H., Rayment, M. D. and van der Bijl, G. (2000) 'An assessment of the total external costs of UK agriculture', *Agricultural Systems*, vol 65, pp113–136

Ritchie, H. (1998) 'A revolution in urban agriculture', *Soils and Health*, vol 57, no 3 Autumn

Röling, N. G. and Jiggins, J. (1998) 'The ecological knowledge system', in Röling, N. G. and Wagemakers, M. A. E. (eds) *Facilitating Sustainable Agriculture*, Cambridge, Cambridge University Press, pp283–307

Rosset, P. M. (1996) 'Cuba: alternative agriculture during crisis', in Thrupp, L. A. (ed) *New Partnerships for Sustainable Agriculture*, World Resources Institute, Washington DC, pp64–74

Rosset, P. (1998) *Eight Myths About Technology and Agricultural Development*. Edited notes from presentation given at the Sustainable Agriculture Forum (SAF), Vientiane, LAO PDR, NCA, 18 June 1998

Rosset, P. M. (2000) 'Cuba: a successful case study of sustainable agriculture', in Magdoff, F., Foster, J. B. and Buttel, F. H. (eds) *Hungry for Profit: The Agribusiness Threat to Farmers, Food and the Environment*, New York, Monthly Review Press, pp203–213

Rosset, P. and Benjamin, M. (1993) *Two Steps Forward, One Step Backward: Cuba's Nationwide Experiment with Organic Agriculture*, San Francisco, Global Exchange

Rosset, P. and Benjamin, M. (1994) *The Greening of the Revolution. Cuba's Experiment with Organic Farming*, Melbourne, Ocean Press

Rosset, P. and Moore, M. (1997) 'Food security and local production of biopesticides in Cuba', *LEISA Magazine*, vol 13, no 4 (December), pp18–19

Scialabba, N. and Hattam, C. (2002) *Organic Agriculture, Environment and Food Security*, Environment and Natural Resources Series No. 4, Rome, FAO

Simon, J. (1997) 'An organic coup in Cuba', *The Amicus Journal*, (Winter 1997), p39

Sinclair, M. and Thompson, M. (2001) *Cuba Going Against the Grain: Agricultural Crisis and Transformation*, Oxfam America. http://www.oxfamamerica.org/newsandpublications/publications/research_reports/art1164.html accessed May 2002

Sligh, R. and Christman, C. (2007) 'Organic agriculture and access to food', Issue Paper Document, *Organic Agriculture and Food Security*, Rome, FAO, pp1–15

Snyder, B. (2003) *Cuba: a Clue to Our Future?* Oakland CA, Food First/Institute for Food and Development Policy

Tansey, G. and Worsley, T. (1995) *The Food System*, London, Earthscan

Vandermeer, J., Carney, J., Gersper, P., Perfecto, I. and Rosset, P. (1993) 'Cuba and the dilemma of modern agriculture', *Agriculture and Human Values*, vol 10, no 3, pp3–8

Weaver, M. (1997) 'Allotments of Resistance', *Cuba Sí* (Summer 1997), p21

Wilson, H. and Harris, P. (1996) *Needs Assessment of Urban Agriculture in Havana*, Coventry, Henry Doubleday Research Association

Ziesemer, J. (2008) *Energy Use in Organic Food Systems*, Rome, FAO

2

Post-Petroleum Food Systems:
Transition and Change

Like a sunset effect, the glories of industrial capitalism may mask the fact that it is poised at a declining horizon of options and possibilities. Just as internal contradictions brought down the Marxist and socialist economies, so do a different set of social and biological forces signal our own possible demise.

(Hawken, 1993)

The challenge of declining oil reserves is one of several to confront humanity in the 21st century. Be it shortages of oil, fresh water or food, ill-health, overpopulation or environmental degradation, these challenges have long been evident and are attributed to modernization and the ensuing disconnectedness between the individual, society and the ecosystems within which we live. In the 21st century more than ever, society is dependent on a range of relatively basic, man-made technologies as a substitute for complex human faculties and natural processes. Using the same economic and industrial structures and technologies that contributed to these challenges, society attempts to address them yet only succeeds in plugging the leaks. This business-as-usual approach is being increasingly questioned, and alternatives provided that demonstrate how to creatively operate an economy within the ecological parameters of the biosphere. Amongst these alternatives, Cuba frequently features as an embodiment of the vision aspired to.

The inevitable decline in fossil fuels and transition to alternatives

Cuba has pre-empted what many industrial countries will experience. By their finite nature, there is no debate over whether fossil fuels – oil, gas and coal – will run out, but rather when this will occur, whether alternative energy supplies will meet the ever-increasing demand, and how best to oversee such a transition. Hubbert, the father of the peak oil concept, predicted that US oil production would peak approximately 30 years after the peak of oil discovery (Hubbert, 1956). He was proved to be correct (Deffeyes, 2006), and with the peak in global discovery occurring in the 1970s, this would mean that the global reserves are, in the first decade of the 21st century, nearing or already past their peak (Mobbs, 2005). In the past few years, a plethora of books have been written on this subject.[1] The majority posit that many countries are now past their oil production peaks and are into the second half of their supplies. According to Strahan (2007), 'There are currently 98 oil producing countries in the world, of which 64 are thought to have passed their geologically imposed production peak, and of those 60 are in terminal production decline.' The rate of decline depends on improvements in extracting the more complicated and costly heavy oils, deepwater oils, polar oils and liquids from gas plants (Campbell, 2005), as well as on the rate of usage, which is linked to oil prices and to the increasing demand from less-industrialized countries. Estimates of oil reserves and improved extraction capacity produced by vested-interest groups, such as oil companies and oil-producing countries, is conflicting, variable and generally unreliable (Hopkins, 2006). Nevertheless, even the major oil companies talk of supply continuing to match current levels of demand for between 20 and 40 years only (Bentley, 2002; Browne, 2005), and all agree on the inevitable transition to alternatives. An analysis of ways to mitigate and manage the oil decline, commissioned by the US Department of Energy, concluded that a response should be initiated at least one decade in advance of the peaking (Hirsch et al, 2005).

Petroleum-dependent agriculture and food systems

In the 20th century, oil and gas took over from the waning extraction of cheap coal reserves, as the drivers of growth of industry, trade, transport and agriculture in industrialized countries. Coal had already had a major impact on the agricultural sector in the 19th century, enabling a shift from hand and animal-drawn power to stronger equipment made of iron and steel, to steam-engine powered field machinery, and to more suitable means of transport (railways and steamships). This transport provided access to more distant markets and sources of soil fertility inputs, including mineral fertilizers (Mazoyer and Roudart, 2006). The takeover by oil heralded more efficient and large-scale industrial, mechanized processes – including the powering of irrigation pumps, production of fertilizers, pesticides and herbicides, mechanization for crop production, storage, drying and processing, production of animal feeds and maintenance of animal operations, and the transportation of farm inputs and outputs.[2]

These industrialized farming systems use approximately 10 calories of fossil energy to produce 1 calorie of food energy (Hamer and Anslow, 2008). In the USA, for example, 29 per cent of farm energy use goes on fertilizer production,

25 per cent as diesel fuel to power farm machinery, 18 per cent for electricity in facility operations, 9 per cent on gasoline for farm machinery, 7 per cent for irrigation and 6 per cent for pesticide production (Brown, 2006). In the UK, agriculture accounts for approximately 2.7 per cent of the nation's total energy use (Ho and Ching, 2008). Most of this, 76.2 per cent, goes to the production and transport of synthetic fertilizers, pesticides, machinery, animal feeds and medicines. Nitrogen fertilizer in particular accounts for 53.7 per cent of total energy use. The remaining 23.8 per cent is used on-farm. An estimated 37 per cent of on-farm energy use is for heating (61 per cent of which is in the protected crops sector), 36 per cent as gas and oil for field operations and especially for arable crops, and 28 per cent as electricity for ventilation, refrigeration, lighting and other facility operations (Adams et al, 2007). Thus, although industrialized agriculture takes a relatively small share of total fossil energy supplies of industrialized countries, the sector cannot operate without it.

Industrialized agricultural systems are not, of course, globally prevailing; they have developed in those countries and regions with more dominant secondary, tertiary and quaternary economic sectors (manufacturing, services and information sectors, respectively): that is, Japan, Canada, the USA, Australia, New Zealand and Western Europe, and increasingly Eastern Europe and former Soviet countries, Israel, and the Asian countries of Singapore, South Korea, Taiwan and (the Special Administrative Region of China) Hong Kong. For less-industrialized countries – those whose economies are dependent on primary industries of mining, agriculture and fishing – industrial forms of agriculture are regionally dispersed in the form of plantation or commodity-oriented production and Green Revolution agriculture. The majority of the world's population, meanwhile, continues to depend on non-industrialized (traditional and intentionally organic) agriculture, yet is encouraged to aspire to more fossil–fuel-dependent systems.[3]

In analyses of alternative farming systems, modern organic agriculture[4] generally consumes less energy than industrialized, although more than traditional subsistence farming. Although organic farming may use more machine hours, this is offset by the absence of other fuel and petrochemical uses. Studies report that organic agriculture performs better on a per hectare scale with respect to both direct energy consumption – fuel and oil – and indirect consumption – synthetic fertilizers and pesticides (Scialabba and Hattam, 2002; Ziesemer, 2008). Any increase in labour requirements of organic farming is an employment opportunity. A comparative analysis of energy inputs on long-term trials at the Rodale Institute, USA, found that organic farming systems use 63 per cent of the energy required in industrial systems, largely because of savings in the energy used to synthesize nitrogen fertilizer (Pimentel et al, 2005). In the UK, the Department for Environment, Food and Rural Affairs concluded that organic crops used 25 per cent less energy than industrialized, and in some crops up to 60 per cent (Hamer and Anslow, 2008). Niles et al (2001) also concluded that it takes 6–10 times more energy to produce a tonne of cereals or vegetables by industrialized agriculture than by more sustainable methods. The actual degree of consumption depends on the type of farming approach – some organic farms are more industrialized than others and may be similarly dependent on irrigation, heavy machinery and heated greenhouses, and on the degree to which the organic farm is embedded in the prevailing food system (Hall and Mogyorody, 2001; Ziesemer, 2008). In fact, the organic movement, which in industrialized countries has developed in the context of plentiful fossil energy supplies, is reviewing the means by which it can reduce energy consumption or even become energy self-sufficient or energy exporting (Hamer and Anslow, 2008).

Industrialized agriculture provides the raw materials for the industrial food system. Although these materials comprise only a small part, the food system is dependent on this cheap and consistent supply for transformation into added-value products, through processing, marketing and distribution activities. Energy usage on-farm comprises a large minority of that of the rest of the food system. Taking figures for the USA and the UK again, production comprises 21 per cent of energy use in the US food system, with home refrigeration and preparation using 32 per cent, processing 16 per cent, transport 14 per cent, and packaging and retailing 7 per cent each (Heller and Keoleian, 2000). In the UK, and using barrels of oil as a measure, production and processing combined attribute to approximately 34 per cent of total oil usage in the food system, while home preparation uses 23 per cent, transport and distribution centres use 18 per cent, catering 10 per cent, packaging 7 per cent and retailing 6 per cent (Lucas et al, 2006). In this sense, the industrialized food system – comprising production, processing, transportation and distribution systems – is inherently vulnerable to both shortages and increasing prices of fossil fuels. Whether organic or industrialized, a more localized, short-chain food system, through which produce is sold with minimal packaging, has lower energy needs. For example, comparative studies indicate that the energy use of organic box schemes may be 90 per cent less than if purchasing similar, industrially produced foods from a supermarket (Hamer and Anslow, 2008).

Using the peak-oil challenge as a driving force for change

Business as usual or creative evolution?

The various scenarios and predictions over life after peak oil have been neatly grouped by Hopkins (2006) into four categories. The first category is one of societal collapse, whereby military force is required to secure remaining hydro-carbon reserves, and chaos ensues. The second category holds that science and technology will find a suitable set of energy replacements, or magic elixir, to continue with business-as-usual. The third category, and the one deemed most likely by Hopkins, is of an enlightened evolution or transition to 'a future sce-nario in which humanity has successfully adapted to the declining net fossil energy availability and has become more localised and self-reliant' (Hopkins, 2006, p19). The fourth and final category describes a more extreme and enforced, post-crisis societal reconstructing toward a more tribal, localized form of community living.

Assuming general agreement on the relative undesirability of categories one and four, this leaves the pursuit of options two (business-as-usual) or three (an enlightened transition). The business-as-usual model assumes that renewable energy replacements, cold fusion or a form of free energy will enable a smooth continuation of current or higher energy consumption. Critics argue that this places faith in so-far undeveloped technology and on top of this assumes pub-lic-sector investment as well as improved, equitable access to this energy for all segments of society. An enlightened transition, on the other hand, is a creatively planned energy descent to reduce per-capita consumption – including through carbon rationing – as well as a transfer to renewables. The strategy would vary by bio-region. Whilst the former model allows for a continuation of current societal structures and power bases, the latter suggests that society wakes up collectively to the real scale of the crisis. Hopkins (2006, p22) notes that whereas the business-as-usual model continues to allow technologies to fix problems, the enlightened transition model requires 'our collective cognitive and behavioural evolution or adaptation as a species'.

It would be more easy to go with the business-as-usual option, and there is a chance that this will work; that sufficient renewable energy replacements will emerge as and when required to meet the growing global demand. Here though, a reminder is made of the preface to this book. The preface explained how the motivation to find out what was happening in Cuba arose not as a result of the peak-oil debate, but because of other, increasingly evidence-based ineffective-ness of the industrialized farming and food system. Whilst the enlightened transition model may appear more risky, and idealistic, it has, largely, to work.

Evidence of the ineffectiveness of the industrial farming and food system

Although successful in enabling the availability of a narrow range of foods to a wide range of people, industrialized and global farming and food systems have a long way to go to be acceptably efficient and effective. They degrade the very natural resource base upon which agriculture (and human life) depends – from soil and water quality to plant DNA structure – and inadequately provide in terms of both quantity and quality.

Degradation of the natural resource base

There is little contestation over the degradation of the natural resource base. Industrial practices result in vast tracts of degraded land, yield declines, loss of plant and animal species diversity, increase in susceptibility to disease, and other serious side-effects over the medium to long term, and have led to a loss of livelihoods (Tansey and Worsley, 1995; FAO, 1997; Conway, 1998; Pingali and Rosegrant, 1998; Oldeman, 1999; Sustain, 2003; Hole et al, 2005). This is particularly so for marginal lands, where the poor soils cannot sustain mono-cultures of annual crops, and which are more vulnerable to flood and drought (Hazell and Garrett, 2001; McNeely and Sherr, 2001). Environmental degradation is also expensive: even a decade ago, agricultural losses due to land degradation were about $550 million annually (Tansey and Worsley, 1995), and the UN estimates that global income loss due to desertification is $42 billion.

Ecologically based, organic farming practices show themselves to be more successful at supporting a broad and adapted diversity of crop species and varieties, building soil fertility and plant resistance to disease and infection, and maintaining clean water courses (Greene and Kremen, 2003; SAN, 2003; Marriot and Wander, 2006). Strengthening the natural resource base also enables farms to better withstand external shocks and stresses, including drought and flood (Holt-Giménez, 2002; Lotter et al, 2003; Ching, 2004). Agriculture accounts for 70 per cent of freshwater use globally, and the UN predicts that, by 2025, 38 per cent of the population will have insufficient water supply (compared with 8 per cent in 2008) (Lang, 2008). Organic practices increase water retention capacity and efficiency by improving soil structure and increasing soil life, by cultivating climatically adapted varieties, and by growing polycultures of deeprooting and ground-covering crops.

Evidence also indicates that organic farming approaches produce lower greenhouse-gas emissions. The reasons for this are threefold: they avoid ammonium nitrate fertilizer (the production of which was responsible for 10 per cent of Europe's industrial gas emissions in 2003), they encourage carbon sequestration through cultivation of deeprooting plants, and livestock's methane emissions are lower if they are feeding on legume pasture (Hamer and Anslow, 2008).

Degradation of the human resource base

Comparing the human health impacts of foods produced under different farming regimes is challenging: the human body is complex, making holistic and generational studies expensive and impractical, scientific paradigms and belief systems differ in terms of sets of specific health indicators and ways of measuring them,[5] and there a few public funds available for research (Heaton, 2001; Niggli et al, 2007; IFOAM, 2008a). Nevertheless, there is no body of evidence to show any human health advantages of consuming industrially produced foods. Conversely, there is a substantial and still-growing body of evidence demonstrating the converse: that ecologically based, organic foods contain more desirable components and fewer harmful ones.[6]

The practice of industrial agriculture has led to a dramatic decline in the macro- and micro-nutrient content of foodstuffs over the last century. For example, mineral levels of fruits and vegetables in the UK have fallen by up to 76 per cent between 1940 and 1991 (McCance and Widdowson, 1940–1991; Mayer, 1997; Baker, 2001), and a similar trend has been seen in the USA and Canada (Rees and Wackernagel, 1996; Bergner, 1997; Davis et al, 2005). This decline is attributed to the unintentional selecting-out of high-nutrient crop varieties when breeding crops for high yield potential, the use of shallow-rooting annuals that are unable to tap into soil nutrients at deeper levels, and the failure to return a full complement of nutrients to the topsoil. Industrialized monocultures also reduce varietal and crop diversity in produce destined for the plate. A combination of different species results in a more balanced diet than any one species can provide, in particular if using traditional seeds and breeds (van Rensburg et al, 2004; Johns et al, 2006).

Several studies and reviews have compared the levels of nutritionally relevant compounds in food produced along organic and industrial lines (Diver, 2000; Heaton, 2001; Worthington, 2001; Magkos et al, 2003; Rembialkowska, 2005). Although results are variable, the trend shows organic plant foods to have higher levels of vitamin C, minerals including iron and magnesium, plant-secondary metabolites or phytochemicals (antioxidants), essential amino acids and dry matter content. Organic foods also tend to show lower levels of nitrate and total protein. Organic meat and milk have more fatty acids, vitamin E and beta-carotene (Adriaansen-Tennekes et al, 2005; Ellis et al, 2006; Soil Association, 2007). Other systematic differences in composition have been found, although the relevance of these differences for human health is not yet known. What is known is that organic crops are more nutrient dense, which arguably offsets yield reductions and rebalances the ratio of food quality to quantity.

Plant-secondary metabolites or phytonutrients require highlighting because of their beneficial effects in the treatment of cancer and other illnesses (Plaskett, 1999). These plant compounds are produced as part of their self-defence or resistance to pests and diseases, and levels are reduced when chemical pesticides and fertilizers are employed (Bennett and Rosa, 2006).[7] On a fresh weight basis, organic foods consistently contain at least 10–50 per cent higher concentrations of plant secondary metabolites than in comparable industrially produced foods (Brandt and Mølgaard, 2001, 2006; Kumar et al, 2004; Heinäaho et al, 2006; Amodio and Kader, 2007; Mitchell et al, 2007; Rembialkowska et al, 2007; Toor et al, 2007).

Ingested pesticide residues are found to damage both the structure and functioning of the immune system in animals and humans, and are also implicated in neurotoxicity, the disruption of the endocrine system and carcinogenicity (BSAEM/BSNM, 1995; Faustini, 1996; Repetto and Balinga, 1996; Colosio, 1997; Voccia et al, 1999; Dewailly et al, 2000; Hancock et al, 2008). Tests on certified organic food show a 3–4 times lower risk of containing pesticide residues than industrially produced food, and following an organic diet eliminates exposure to common insecticides (Lu et al, 2005; Winter and Davis, 2006). The handling of pesticides results in 3 million pesticide poisonings a year and 20,000 deaths, largely in less industrialized regions (WHO, 1999).

The majority of food evaluation studies focus on the composition of individual foods and single nutrients, and very few on holistic feeding studies. Three epidemiological studies do indicate improvements in physical, mental and psychic well-being, including increased immunological activity, for humans following organic diets (Alm et al, 1999; Alfvén et al, 2006; Leiber et al, 2006). Studies on animals are easier to undertake, and feeding trials in general indicate significant improvements in the growth, reproductive health and recovery from illness of animals fed organically produced feed (Heaton, 2001; Worthington, 2001; Finamore et al, 2004; Adriannsen-Tennekes et al, 2005; Lauridsen et al, 2005; Baranska et al, 2007).

An ineffective food system

The aim of any food system, one might suppose, would be to feed people with an adequate, or even abundant, supply of nutritious foods. As it happens, this is a human right. As with previous agricultural innovations throughout the centuries, industrial agriculture – and its tropical counterpart, the Green Revolution – has produced significant yield increases. Between 1950 and 1984, world grain production increased by 250 per cent (Kindell and Pimentel, 1994). Yet the industrial food system has so far failed to provide everyone with an adequate diet, even where it has been subsidized (such as in Green Revolution technology packages). In 1996, nations at the World Food Summit pledged to halve the number of undernourished people by 2015. By 2006, the FAO's annual report, The State of Food Insecurity in the World, reported that 'Ten years later, we are confronted with the sad reality that virtually no progress has been made towards that objective' (FAO, 2006, p6). The report continues: 'The world is richer today than it was ten years ago. There is more food available and still more could be produced without excessive upward pressure on prices. The knowledge and resources to reduce hunger are there. What is lacking is sufficient political will to mobilize those resources to the benefit of the hungry.' Between 2001 and 2003, the FAO estimated that there were 820 million undernourished people in developing countries, 25 million in transition countries and 9 million in industrialized countries (FAO, 2006). UNICEF estimates that one-third of the world's population of more than 6 billion are affected by food-related ill health, such as primary nutrient deficiencies and corresponding illnesses, in both industrialized and less-industrialized regions (Baker, 2001; WUR, 2002).

Nonetheless, more ecologically based, organic production approaches are sidestepped by international development agencies and national ministries of agriculture owing to their reportedly low yield performance and, therefore, their apparent inability to meet global food needs or be appropriate in food insecure situations (IAC, 2003). In fact, early yield comparisons between certified organic and industrial agriculture has indicated a yield decline of approximately 20 per cent for organic production. However, these studies were based on the performance of certain market-oriented organic systems in temperate climatic regions. Whereas outputs of any one specific crop may be lower on an organic farm than an industrialized one, total farm yields are higher (Altieri et al, 1998). More recent studies show non-certified organic farming approaches to achieve significant yield increases over both traditional and industrial agriculture, and in particular in resource-poor regions on marginal lands and in tropical and subtropical climates (Pretty and Shaxson, 1997; Souza, 1998; Altieri et al, 1999; McNeely and Scherr, 2001; Mäder et al, 2002; Parrott, 2002; Parrott and Marsden, 2002; Pretty et al, 2002; Rundgren, 2002; Delate and Cambardella, 2004). An analysis by Badgley et al (2007) indicates that organic methods, which use leguminous cover crops to replace nitrogen fertilizer, could produce enough food on a global per capita basis to sustain the current human population, and potentially a larger population, without increasing the agricultural land base. Overall, not only is the common uncontextualized focus on yield performance over the short term based on outdated evidence, but it also interferes with achieving food security goals. This focus diverts attention from equally important goals of guaranteeing harvests, increasing community resilience to shocks and stresses, and enabling local availability of a diverse range of quality foods (Bindraban et al, 1999; Wright, 2005). Moreover, there is a strong case that if ecologically based, organic systems had a fraction of the investment poured, at taxpayers' expense, into industrial farming, their performance would be greatly enhanced (Pretty et al, 1996).

Within the food security debate generally, issues of food quality and diversity have been overlooked. The industrial food system relies on the supply of cheap raw materials to be transformed into added-value products. It actually offers a very narrow nutritional diversity, being based on a core group of about 100 basic food items which consist of 75 per cent of our food intake (Lang and Heasman, 2004). In the UK, for example, 80 per cent of food is processed, and the food industry spends over $1,200 million annually on advertisements to promote these products. In contrast, the budget for advice on health education by the UK Health Education Authority is about $4 million a year, and excess consumption of fat and under-consumption of fruit and vegetables are major factors in the nation's top two causes of premature death: coronary heart disease and cancers (Lang, 1997). Obesity in less industrialized regions is also rising, mainly in urban areas with Western diets, and in rural areas where new technologies reduce the need for physical activity (FAO, 2002). In industrially processed foods, over 500 additives are permitted for use, and have been linked to allergic reactions, headaches, asthma, growth retardation and heart disease. The organic processing sector permits around 50 additives, and also regulates to avoid antioxidant-suppressing and high-temperature technologies (Heaton, 2001; Benbrook, 2005). Another overlooked issue of food quality is that of food deterioration. Foods consumed as soon after harvest as possible retain more of their nutritional value. Additionally, beneficial components such as antioxidants do not last long once ingested and need to be consumed at most meals in order to sustain optimal levels in the body (Benbrook, 2005). Fresh food sources such as fruits and vegetables are superior to the consumption of equivalent vitamin and mineral supplements in terms of nutritional factors such as antioxidant activity (Dragsted et al, 2004).

Further inefficiencies in the food system comprise food losses, food wastage and the externalized costs of environmental and human health impacts. In terms of food losses, and if chemical preservatives are avoided, several studies show unsprayed organic produce (such as wheat, apples, potatoes) to have lower storage losses and significantly lower disease scores than unsprayed non-organic (Raupp, 1996; Granstedt and Kjellenberg, 1997; Birzele et al, 2002; Pedersen and Bertelsen, 2002; Moreira et al, 2003). This is thought to be due to natural plant toxins that suppress rots and moulds (Benbrook, 2005). In terms of wastage, industrialized nations throw away more than half the food produced each year. In the UK this discarded amount – 20 million tons annually – is valued at $40 billion and equivalent to half of the food import needs for the whole of Africa. Approximately three-quarters of wastage occurs in homes, shops, restaurants, hotels and food manufacturing. The rest is lost between the farm and the shop shelf (Mesure, 2008). In addition to the costs of cleaning up environmental damage, the restoration of human health is also a hidden cost of the food system that is not included in the cost of food, but instead is paid for by governments and society. In 1996, these health costs amounted to $81–117 per hectare in Germany and $343 per hectare in the UK (Pretty et al, 2000). The ineffectiveness of the food system also shows up in relation to income inequities. Farming accounts for half the world's workforce, yet is one of the most poorly paid professions in the world, with thousands of farmers going out of business or committing suicide annually (Lang, 2008). Rising oil prices are affecting the cost of food, but prior to this farm gate prices tended to fall or remain fairly constant, whereas profits in the food industry soared. In 2005, for example, the combined profit of the world's top five food and beverage manufacturers was $33 billion (Oram, 2008). Despite its power and influence, the food industry has remained largely unaccountable in the food security debate (Wright, 2005).

The static state of agricultural development

In short, the industrialized farming and food system has not fully delivered, and to date neither has the tinkering around the edges with 'more sustainable' agriculture. Evidence indicates that smaller-scale, localized and ecologically based organic production would be more efficient in terms of increasing availability of nutritious foods, minimizing post-harvest losses, providing intelligent yield increases and internalizing costs. Given this, it would seem logical to shore up the food system to reduce wastage, increase nutritional value, restore degraded lands, intensify adaptive, sustainable production, ensure equitable access through localized production, and so on, rather than continue to assume that a new technology will provide the solution. For it is not as if practising industrial agriculture will lead us to a future situation where we can relax about food supplies and convert to more sustainable organic production. As described above, quite the reverse happens. The fact that this evidence has not already obligated widespread change indicates that other forces are at play to maintain the current system as it is.

Rather than decisions being based purely on evidence, people have differing perceptions and sets of beliefs, and this rests in the concept of constructivism. Although there may be a given set of universal truths, each person or group constructs and continually reconstructs its own reality. Such construction is based on sense-making processes – a set of cognitive values, beliefs, perceptions, emotions and accrued experiences – and it is these with which the individual, or group, observes, understands and learns (Kolb, 1984; Kloppenburg, 1991). Thus evidence may be interpreted in various ways, or ignored if it conflicts with strongly held beliefs or values. Constructivism helps to explain the evolutionary progress of agriculture over the last half century, or rather the lack of progress. In Europe following the Second World War, a justifiable call to rebuild and shore up food reserves saw the mainstreaming of large-scale, industrial agriculture in many regions (Conford, 2001). When targets had been achieved and the pressure reduced, and as evidence emerged in the 1960s on the negative impacts of these practices, an evolutionary learning approach would have led to more sustainable agricultural systems. Yet this did not occur to any substantial degree, even though food insecurity in these European countries had been almost eliminated. What has maintained and preserved agriculture in a static industrialized state, given the rapid changes in other sectors of society? The post-war growth of a flourishing agribusiness structure to provide the inputs, equipment and other goods for industrial agriculture comprises a part, but not necessarily all, of the explanation. Agribusiness, if not operating out of an alternative economic construct of beliefs or values, has the power to maintain the industrialized production system in order for it to continue expanding sales, keeping production costs down, increasing profits and maintaining jobs (Vorley, 2003). As such, a natural process of evolutionary change and improvement in the sector has not occurred.

To conclude, to respond to the peak oil challenge with a business-as-usual model would maintain the global food system as is, but powered by renewables. Yet even if this gamble pays off and society is able to continue without strolling headlong into a sharp energy shock, evidence indicates that other crises will occur within this industrialized model, which makes logical the suggestion to use peak oil as a wake-up call for more creative change. In fact, Hopkins (2006, p19) notes that energy descent 'is a term favoured by people looking toward energy peak as an opportunity for positive change rather than an inevitable disaster'. The constructivist concept highlights that, rather than there being a right or wrong way, society is broadly able to create whatever system it believes in and values.

Transition and change to a post-petroleum farming and food system

The purposeful design and implementation of post-petroleum systems necessitates a process of transition. Amongst development policy-makers, there is little consensus over a strategy for stimulating localized, small-scale agriculture in less industrialized countries, and a lack of vision on how to build nutritional security other than make modifications to existing systems (Farm Africa, 2004; IFOAM, 2008b). In most countries, generally, intentional organic production is spatially fragmented, and knowledge on its development mechanisms and support needs are, as a result, difficult to extrapolate. Social movements for change are pioneering the design and development of localized food systems, yet these also struggle to identify practical examples to draw from, and consistently turn to Cuba, which has the reputation for being the only country with a policy commitment to a nationwide localized and organic production approach.

Characteristics of a post-petroleum food and farming system: localized and organic

What would a farming and food system look like, one that is both based on renewable energy and is more energy efficient, one that might be in operation in Cuba? If it is to deliver on food needs, the changes in energy source and quantity would necessitate the scaling-up of more localized and organic systems. The concept of localization is characterized by community self-reliance as far as practicable, diversified local economies, and living in harmony with nature (Schuman, 2000; Norberg-Hodge, 2002; Hopkins, 2006). According to Feenstra (2002), a local food system is 'a collaborative effort to build more locally based, self-reliant food economies – one in which sustainable food production, processing, distribution and consumption is integrated to enhance the economic, environmental and social health of a particular place'. Kloppenburg et al (1996) identify a foodshed or bio-region as being the basic ecological unit from where local food is sourced. This foodshed is defined by climate, soil watershed, species and agro-ecosystem, and its size varies depending on the availability of year-round foods and the variety of foods grown and processed. Local is a relative term; in sparsely populated or flat regions the local area may be larger than those of dense communities or mountains. Components of a localized food system include direct farm sales and farm shops, Community Supported Agriculture (CSA) farms, farmers' markets, box schemes, community gardens, food cooperatives and local retailers. New retail forms would undoubtedly develop if localized systems were mainstreamed, as current players in the food retail system struggle to find a new role for themselves. Local produce would be defined by its seasonal availability; local, heritage varieties and recipes; freshness and flavour; and (acceptable) variability in produce appearance.

Agriculture in a post-petroleum society would necessitate organic alternatives to petrochemical inputs. The move to more localized systems would further drive organic growth. Farms producing for local markets would have to shift from producing large quantities of a small range of produce to smaller quantities of a more diverse range in terms of both crops/livestock and varieties. This in turn would increase farm complexity and would provide opportunities for increased levels of recycling of farm components and for establishing and maintaining an ecologically balanced system. More localized production would also spawn wider consumer interest and concern over the way that their food is produced, as a sense of community develops.

The widespread uptake of ecologically based, organic farming would see changes in land use and human resources. In order to achieve sustainable stocking rates that can be maintained on homegrown or locally grown feed, some of the land currently put to intensive meat and dairy production would need to be freed up for fodder crops. To improve soil fertility, more land would be put to nitrogen-fixing crops, and human sewage recycled as fertilizer (Harnapp, 1988; Offerman and Nieberg, 2000; Fairlie, 2007). This in turn would necessitate cultural changes to reduce the meat content of diets. Given the biophysical limitations on the capacity to expand and specialize, the average size of land holding would decrease and farm numbers would increase (Campbell and Coombes, 1999). Farm labour would also generally increase, depending on the type of production system. For example, a temperate mixed farm which included on-farm processing and direct marketing would have labour increases of approximately 20 per cent (Offerman and Nieberg, 2000). These increases would require land reform; they would be reversing the trend in industrialized countries, where, for example, the number of farms in the UK have fallen between the 1930s and the present, from 500,000 to 130,000, and 70 per cent of land is owned by less than 1 per cent of the population. In the USA, farm numbers have fallen from 6.5 million to under 2 million over the same period (Vidal, 2003). As well as a general increase in the number of agricultural courses and students, specialist skills in ecological knowledge, farm management and local adapatability would require strengthening, these skills previously having been substituted by agrochemical inputs and simplified production routines (NAS, 1989; Pretty, 1995). The key characteristics or indicators of a change from industrialized to widespread organic farming are listed as follows (Mollison, 1988; Sattler and Wistinghausen, 1992; Altieri, 1995; IFOAM, 1998):

- increase in production of vegetable protein and decrease in animal protein
- increase in land put to legume and green manure crops, and decrease in pastures and grazing
- elimination of external inputs of chemical fertilizers and pesticides, and significant decrease in feed concentrate
- increase in mixed farming and mixed cropping
- expanded knowledge base in local ecology, biology and environment
- reliance on the skills of observation, ecological literacy and good management practices
- pest and disease control by prevention rather than cure
- increase in labour requirements
- increase in the use of adapted traditional techniques and local knowledge
- increase in local innovation
- production principles based on spatial and temporal diversity, optimal use of resources, water conservation, controlled succession
- presence of wild areas
- productivity optimized over the long term rather than maximized over the short term

- use and re-use of local resources
- emphasis on building farm complexity and throughputs rather than single commodity outputs
- intelligent use of scientific knowledge and appropriate technologies
- avoidance or absence of techniques that imbalance, destroy or alter life forms, deplete resources or work against ecological principles.

Based on empirical models, the transition from industrialized to organic farming comprises three to four main stages, culminating in a designed state of a self-regulating farm organism (Mollison, 1988; MacRae et al, 1990; Altieri, 1995; Pretty, 1998). From the starting point of an industrialized monoculture, the first step in a transition is to reduce the intensity of agrochemical application through more rational usage, and to introduce organic inputs and integrated management strategies. The second step comprises a greater use of organic inputs, and an increase in farm efficiency, precision techniques and recycling. Yields may reduce during this period, whilst biodiversity starts to increase. Moving on to step three entails the development and expansion of ecological knowledge in order to reduce the need for any kind of input substitution and introduce regenerative technologies. Here, labour demand increases, input costs drop and economic performance rises. The final stage in the transition process comprises a system redesign based on ecological principles in order to create a balanced farm entity. This design may include an increase in perennials and no-till methods, as well as more social and community activities and interactions. The timescale for this transition period has been estimated at between three to eight years, depending on the design vision. Although the final stage is seen as the most desirable in terms of sustainability, earlier stages are more popular with policy-makers as input substitution is cheaper to implement.

This transition is accompanied by a change in attitude, which is seen by some as a key prerequisite to any substantial change in practice (such as Lampkin, 1990; Goewie, 2002a). Others suggest that a change in behaviour and practice may then induce a change in attitude (Pretty, 1998; Hall and Mogyorody, 2001). For a planned energy descent, an attitude change would precede, whereas a crisis, as happened in Cuba, would probably force practical change first.

The food system would also need to localize, with short production–consumption changes and far greater local complexity. To address food security needs, characteristics would be required that enable the achievement of long-term food availability and accessibility as well as of nutritional adequacy. These are listed below (Wright, 2005):

Achievement of long-term food availability:
- high total yields through intensified intercropping and polycultures

- maximization of land under cultivation through zero-degradation and the regeneration of marginal lands
- broadened availability of food types through diversified production and lengthened cropping seasons
- zero postharvest food losses through increases in local processing, promotion of fresh food consumption (including non-uniform foods) and shortened market chains
- consumer education on seasonal limitations to availability.

Achievement of long-term food access:
- localized and short production–consumption chains
- affordable prices and food exchange mechanisms
- increased numbers of people involved with, and with direct links to, production
- more complex food linkages and distribution networks
- food price related to proximity of production.

Achievement of long-term food nutritional adequacy:
- strengthened links between food production, preservation and consumption, both within households (home-grown) and between households and producers
- education on the relationship between food production and nutrition
- improvements in home and local storage facilities
- increased consumption of a diverse range of fresh produce.

Policy and institutional mechanisms to support local and organic food systems

To date, organic and localized systems have occurred often in the face of prevailing policy and institutional arrangements, rather than because of them. There is little knowledge of what works in terms of sustained institutional support from government policy, legal and land tenure structures, research and extension services, the private sector or markets, nor of the way in which the farmer – or other food producer – engages with these groups. If the structure of food production is to change significantly, this has major implications for the human networks that exist around it (Röling and Wagemakers, 1998). These networks provide employment far in excess of the number of primary producers, and yet they depend on the continuation of the underlying farming structure (MacRae et al, 1993). Thus it is a sensitive process to make changes to the agricultural system that might upset this huge institutional support system, and especially those that might require changes in composition, skill sets or power bases.

For centuries, rural farming communities operated their own, localized support systems, involving the exchange of knowledge, goods and services. With the onset of industrialization, these aspects became externalized, so that farmers were perceived as clients for the receipt or transfer of technology that was developed in the research field or laboratory. Industrial science is concerned with knowledge emanating from controlled and repeatable experiments, and by causal relationships between action and reaction. The outcomes are sets of products to sell to the farmer as inputs, with the same product being applied over a large area. Compared with this, organic science is concerned with processes, such as nutrient and energy cycles, decomposition and succession. It takes a holistic approach, seeing the component parts as forming a greater whole (Wolfert, 2002). The outcomes of this science are techniques to manage and work with nature more effectively, and as such need to be more location-specific and need to involve the farmer in the research process. The more holistic the approach, the fewer saleable outputs. Organic knowledge itself arises from the use of cognitive, intuitive and experiential learning through interacting with the natural environment on an ongoing basis (Goewie, 2002b; Wright, 2004), and the acquisition of this knowledge has been termed 'ecological literacy' (Pretty, 2002). The science of ecology is used to explain and understand rather than to drive agriculture. In this sense, the paradigms and methodologies that form the basis of agricultural research, development and innovation for organic agriculture are quite different from industrial agriculture.

The institutional and policy drivers of farming and food systems comprise degrees of government regulation, markets and social negotiation and agreement (Smith, 1989; Hood, 1998). In industrialized countries, the market driver is most dominant, whereby prices are linked to supply and demand. Where there is sufficient or over-supply, and where the market is unable or uninterested to differentiate on quality, free-market pressure drives down product prices and forces farmers to follow the latest cost-saving (and frequently unsustainable) technological trend in order to compete, to prevent themselves going out of business. This relentless cycle over seasons and generations is termed the 'treadmill effect' and has led to unsustainable farming practices and farming livelihoods (Cochrane, 1958; Röling, 2002). In contemporary certified organic systems, social negotiation takes place whereby consumers agree to pay a higher price to enable farmers to internalize costs and therefore farm more sustainably. For a more widespread organic farming and food system, driving mechanisms may require a greater degree of regulation as well as of social interaction. Other policy and institutional mechanisms to support local organic farming systems have been suggested as follows (Pretty, 1998; Mansvelt and Lubbe, 1999; Heinberg 2007):

• location-specific management and policies

- increase in participatory research, extension and other mechanisms for social engagement
- focus on knowledge and learning over technology diffusion
- emphasis on managing agro-ecosystems instead of applying external inputs
- decentralization of institutions
- tendency away from the free market as the dominant driving mechanism
- policy support for organic practices, and reduced support or increased regulation for industrialized practices
- integration of holistic and systems principles
- inclusion of indigenous and informal knowledge bases in decision-making
- land reform
- loans and other financial incentives
- higher and stabilized food prices
- revitalization of farming communities and farming culture.

Facilitating change to localized systems

Whether because of a sudden crisis or a planned transition, people have to cope – that is, they make relevant adaptations to the environment as they perceive it. This coping may be passive, active or creative (Huang, 1998). Coping with an energy crisis, for example, may take the form of ignoring it or, at the other extreme, completely reorganizing the whole system. The basic coping process is one of learning, a cycle of observation, of understanding what has been observed, of thinking and reflecting on this, and then of acting upon it (Kolb, 1984). At the level of society, this learning process results in transitions over the long term, and the transition process may be better supported if the various stages are recognized and understood (Rotmans et al, 2001).

Pioneering societal groups have for some years been developing initiatives to facilitate change to more localized, carbon-free systems. Common features of these groups are that they are driven by an individual with a vision, they lay out a clear set of sustainability principles, they focus on practical change, and they work against the tide of mainstream thinking and action. A longstanding example is the Bioregional Development Group, based in London, which formed in 1992 to demonstrate the feasibility of 'One Planet Living': maintaining a high quality of life whilst reducing wasteful consumption and using a fairer share of the Earth's resources. The Group advocates for local production and recycling to reduce transport, fossil-fuel use and CO^2 emissions, produces commercially viable products and services, and promotes a set of principles that include local resource-use efficiency, closing the loop, using appropriate scale technology, network production (producing locally with centralized coordination and marketing) and fair trade (Desai and Riddlestone, 2002).

A later example of transition management is the Transition Towns Initiative, which sets out and supports a clear process of how to go about achieving local, low-energy, resilient economies in response to the twin pressures of peak oil and climate change. Emerging from the UK in the mid-2000s, it uses the concept of 'backcasting' – that is, imagining a positive future scenario and then working back to identify how this vision had been (or could be) achieved (Cook, 2004). To avoid the problem of a lack of consensus on the desired vision, the Transition Initiative approach backcasts, based on agreed sustainability principles (Hopkins, 2006). To do this, it asks questions such as 'What would the community [farm/food system] look like if it was emitting less CO^2 and using less non renewable energy and was well on its way to rebuilding resilience in all critical aspects of life?' (Brangwyn and Hopkins, 2008, p29). To enable a community to work through the transition process to achieve their sustainability goals, specific steps are recommended. These steps place an emphasis on community self-organization and inclusiveness, and include identifying and valuing local resources and traditional knowledge, building skills that can make a practical contribution, and working alongside local policy-makers (Hopkins, 2007). Institutions are involved as stakeholders, and at the same time more sustainable infrastructure is built – termed parallel public infrastructure – which complements existing infrastructure but can take over if and when the existing one fails or does not change (Darley et al, 2006; Hopkins, 2006).

Both these initiatives provide examples of the emerging alternatives to the peak-oil challenge, promoting an enlightened transition to a post-petroleum food system. Both circumvent the challenges of lack of consensus and of institutional reticence to change. At 2008, over 50 towns in the UK were officially in transition and several hundred more preparing to do so including in other countries (Transition Towns, 2008). The Transition Towns Initiative, and others, use the Cuban experience as a role model to aim for (Pfeiffer, 2006; Heinberg, 2007).

Understanding Cuba's adapted food and farming system and the way it developed holds relevance from several angles. Even in the face of a crisis, could Cuba have made a transition to a localized and organic food system, one that fed the population? If so, what were the factors that supported this, and if it had not made this transition, then why not? As well as providing a learning example for the peak-oil challenge, these insights have broad applicability for other resource-poor countries attempting to increase their food security status.

Notes

1   See, for example, Campbell, 2005; McKillop, 2005; Simmons, 2005; Deffeyes, 2006; Heinberg, 2007; Tertzakian, 2007.

2   In the case of Europe, the emergence of industrialized agriculture came with the industrialized production of chemical fertilizers (based on the chemist von Liebig's

rejection of humus in favour of the use of chemical components as a plant growth medium) and the rapid spread of urbanization in the 19th century. (Specifically for Britain, major additional triggers were the Repeal of the Corn Laws in 1846, which pushed through free trade and a decline in British agriculture, and after the Second World War the drive for mass food production and related consumerism: Conford, 2001.) The dominant agricultural approach of the 20th century in industrialized countries relies upon manufactured pest and disease controls and fertilizers, and emphasizes the maximizing of production through simplification, the use of external technologies and minimizing labour requirements (Pretty, 1998). Because this non-conventional form of agriculture came of age with the Industrial Revolution and the technological possibilities emerging therefrom, because it was based on the Revolution's underlying concepts of standardized, mass production and consumption, capital accumulation, and reliance on large amounts of fossil energy and industrially produced inputs, so it was termed industrialized (Kaltoft, 2001). Production approaches falling under this category include integrated, precision, high-tech and certain 'sustainable' definitions (Goewie, 2002a). An industrialized system is associated with socio-economic issues of external dependency, long marketing chains, cost externalizations, and free-market principles as a driving force. Guiding and driving all this is a particular set of attitudes and perspectives surrounding agriculture, such as the belief that mankind can break free from and take control over the natural environment and natural processes, and that this is a positive step. The development of GM crops is a contemporary example of this belief (Wright, 2005).

3   Although traditional, ecologically oriented forms of agriculture have been ubiquitous since the emergence of agrarian communities, a more clearly defined organic agriculture, and the organic movement, emerged as a coherent entity once a sizeable alternative had presented itself. The early organic movement grew in opposition to the industrialized approach, with several key areas of disagreement. Prophetically, these areas included the degenerative effects of industrial agriculture on human health and in particular the rural and urban poor (Picton, 1943; McCarrison, 1945; Wrench, 1972); the loss of rural community and rural unemployment (Hyems, 1952); the loss of soil humus, soil life, and the recapturing and recycling of nutrients (Howard, 1940; Balfour, 1943); the loss of biodiversity and variety (Stapledon, 1942); and the potential danger of substituting new scientific findings for accumulated experience and ecological principles (Massingham, 1942; Stapledon, 1942). The industrialized model – of competitive yield maximization over the short term, was held as incompatible with an organic model of collaborative optimization of outputs and synergies over the long term. It is important to note that the early organic movement was concerned with agrochemicals because of the mindset they represented rather than purely because of their potentially toxic properties (Conford, 2001), and the roots of organic agriculture came about before the use of synthetic fertilizers was large enough to become an issue (Woodward, 1999). In other parts of the world, a major trigger for stimulating an alternative, organic agriculture has been the experience of the negative impacts of Green Revolution technologies, such as in parts of India. Negative impacts of Green Revolution technologies identified by Hazell (1995) include the destruction of beneficial insects, waterlogging and salinization of irrigated land, pollution of groundwater and rivers, poisoning of farm workers, and excessive dependency on a few improved crop varieties. Sinha (1997) describes how, in India, the realization of the destruction

of the agro-ecosystem by Green Revolution techniques has led farmers and scientists to develop alternatives – 'perhaps a non-chemical agriculture', which includes the reviving of traditional techniques and identifying cheaper and ecologically safer alternatives to agrochemicals, such as biofertilizers and biocontrols, sewage farming and saline agriculture, and systems that produce more nutritious foods with shorter harvest cycles. In these less-industrialized countries, the socio-economic inconsistency of industrialized agriculture with local, community-based systems is perhaps more pronounced than for industrialized countries (Crucefix, 1998; Bakker, 2003). Other driving forces, or triggers, for change to organic methods are for commercial, health, quality of life, environmental, religious, philosophical and moral reasons, and are seen as much in less-industrialized countries as in industrialized (Fernándes et al, 1997; Merlo, 1997; Anandkumar, 1998; Harris, 1998; Hall and Mogyorody, 2001). Unifying these cases is a conscious decision to turn away from one approach and towards another. It is this conscious decision that signifies the turning point between traditional and organic agriculture; the former being 'by default' and the latter being intentional.

4   Note that the term 'organic' in this book means an ecologically oriented farming approach, using regenerative husbandry techniques based on ecological principles and science, without the market and standards connotations of certified organic agriculture or the unqualified knowledge and input levels of 'sustainable' agriculture. Three core management strategies support this complex relationship (Mansvelt and Lubbe, 1999): 1) maintaining semi-closed and integrated systems (including high levels of recycling of components); 2) supporting natural balance through diversity in time and space; and 3) respecting and encouraging integrity both of the organism to fulfil its inherent potential, and of man's position with respect to nature. Utilizing both traditional and scientific knowledge as well as common sense, intuition and ethical judgement, the organic mindset is rooted in intergenerationally successful forms of traditional agriculture, and in this broad sense encompasses like-minded and synonymous farming approaches such as biodynamic, natural farming and agro-ecology. In this definition, ecological literacy and intent are the major indicators of an organic system.

5   For example, more holistic methods such as biocrystallization and biophoton emissions have been developed within the organic research sector to test for properties of organic foods, but these are not as yet recognized by the mainstream (Anderson, 2001).

6   The concept of health is central to organic production. Health is defined not simply as the absence of illness, but the maintenance of physical, mental, social and ecological well-being (IFOAM, 2008a). This has guided the development of organic farming practices, to avoid those which may have adverse health effects. The cornerstone of health in an organic system is the soil, through a 'living soil' that is acknowledged as having the capacity to influence and transmit health through the food chain (Woodward, 2001).

7   In nature, plants depend on nutrients released from the decomposition of organic matter. If plants grow relatively slowly, they build up their chemical defences to a level that prevents most pests and diseases but that does not harm humans, because the body can metabolize or excrete them without ill effect (Friedman, 2006). However, if a plant is allowed to grow unusually fast by providing it with an abundance of

nutrients such as through agrochemicals, the accumulation of defence compounds is reduced (Stamp, 2003), and this reduction also occurs if insecticides and fungicides are applied, as the plant's own natural defence mechanisms are not activated. Plant breeding and the selection-out of distasteful compounds also reduces the level of natural plant toxins including phytonutrients.

References

Adams, S., Langton, A. and Plackett, C. (2007) 'Energy use in agriculture and horticulture; moving to a low carbon economy', *The Commercial Grower* (9 August). http://www2.warwick.ac.uk/fac/sci/whri/research/climatechange/cgenergy/ accessed August 2008

Adriaansen-Tennekes, R., Bloksma, J., Huber, M. A. S., de Wit, J., Baars, E. W. and Baars, T. (2005) *Organic Products and Health. Results of Milk Research 2005*, The Netherlands, Louis Bolk Instituut Publications GVV06

Alfvén, T., Braun-Fahländer, C., Brunekreef, B., von Mutius, E., Riedler, J., Scheynius, A., van Hage, M., Wickman, M., Benz, M. R., Buddle, J., Michels, K. B., Schram, D., Ublagger, E., Waser, M. and Pershagen, G. (2006) 'Allergic diseases and atopic sensitisation in children related to farming and anthroposophic lifestyle. The PARSIFAL study', *Allergy*, vol 61, pp414–421

Alm, J. S., Swartz, J., Lilja, G., Scheynius, A. and Pershagen, G. (1999) 'Atopy of children of families with anthroposophic lifestyle', *Lancet*, vol 353, pp1485–1488

Altieri, M. (1995) *Agroecology*, second edition, London, IT Publications

Altieri, M. A., Rosset, P. and Thrupp, L. A. (1998) *The Potential of Agroecology to Combat Hunger in the Developing World*, 2020 Brief 55 (October), Washington DC

Altieri, M. A., Companioni, N., Canizares, K., Murphy, C., Rosset, P., Bourque, M. and Nicholls, C. I. (1999) 'The greening of the "barrios": urban agriculture for food security in Cuba', *Agriculture and Human Values*, vol 16, pp131–140

Amodio, M. and Kader, A. (2007) Unpublished data. University of California, Davis

Anandkumar, S. (1998) 'Motivating farmers to convert to organic farming and strategies for organic extension', *Ecology and Farming*, no 18 (May–August), Germany, IFOAM

Anderson, J.-O. (2001) 'Food quality and health – the need for new concepts and methods', Presentation at Triodos Bank Annual General Meeting, London, 7 April

Badgley, C., Mohhtader, J., Quintero, E., Zakem, E., Chappell, M. J., Aviles-Vazquez, K., Samulon, A. and Perfecto, I. (2007) 'Organic agriculture and the global food supply', *Renewable Agriculture and Food Systems*, vol 22, no 2, pp86–108

Baker, B. (2001) *The Truth About Food*, Bristol, Soil Association

Bakker, N. (2003) Personal communication with Oxfam extension worker, Mozambique. Wageningen

Balfour, E. B. (1943) *The Living Soil*, London, Faber & Faber Ltd

Baranska, A., Skwarlo-Sonta, K., Rembialkowska, E., Brandt, K., Lueck L. and Liefert, C. (2007) 'The effect of short term feeding with organic and conventional diets on selected immune parameters in rat', Paper presented at *3rd QLIF Congress: Improving Sustainability in Organic and Low Input Food Production Systems*, University of Hohenheim, Germany, 20–23 March

Benbrook, C. (2005) 'Breaking the mold – impacts of organic and conventional farming systems on mycotoxins in food and livestock feed', *State of Science Review*, The Organic Center, USA. http://www.organic-center.org/ reportfiles/Mycotoxin_SSR. pdf accessed July 2007

Bennett, R. N. and Rosa, E. A. S. (2006) 'Phytochemicals under organic and low input crop production systems – potential influences on health and nutrition in humans and animals', Paper presented at *Joint Organic Congress*, Odense, Denmark, 30–31 May

Bentley, R. W. (2002) 'Global oil and gas depletion: an overview', *Energy Policy*, vol 30, pp189–205

Bergner, P. (1997) *The Healing Power of Minerals, Special Nutrients and Trace Elements*, Rocklin CA, Prima Publishing

Bindraban, P., Koning, N. and Essers, S. (1999) *Global Food Security. Initial Proposal for a Wageningen Vision on Food Security*, Wageningen University

Birzele, B., Meier, A., Hindorf, H., Kramer, J. and Dehne, H. W. (2002) 'Epidemiology of Fusarium infection and deoxynivalenol content in winter wheat in the Rhineland, Germany', *European Journal of Plant Pathology*, 108, pp667–673

Brandt, K. and Mølgaard, J. P. (2001) 'Organic agriculture: does it enhance or reduce the nutritional value of plant foods?' *Journal of the Science of Food and Agriculture*, vol 81, pp924–931

Brandt, K. and Mølgaard, J. P. (2006) 'Food quality', in Kristiansen, P., Taji A. and Reganold J. (eds) *Advances in Organic Agriculture*, Australia, CSIRO Publishing, pp305–327

Brangwyn, B. and Hopkins, R. (2008) *Transition Initiatives Primer*, Version 25, 11 January 2008, Transition Initiatives. http://transitionnetwork.org/Primer/TransitionInitiatives Primer.pdf

Brown, L. R. (2006) *Plan B 2.0: Rescuing a Planet Under Stress and a Civilisation in Trouble*, New York, W.W. Norton & Co.

Browne, J (2005) Speech to World Petroleum Congress, Johannesburg, South Africa. www.bp.com

BSAEM/BSNM (1995) 'Evidence of adverse effects of pesticides', *Journal of Nutritional and Environmental Medicine*, no 5, pp341–352

Campbell, C. J. (2005) *Oil Crisis*, Brentwood, Multi Science Publishing

Campbell, H. and Coombes, B. (1999) 'Green protectionism and organic food exporting from New Zealand: Crisis experiments in the breakdown of Fordist trade and agricultural policies', *Rural Sociology*, vol 64, no 2, pp302–319

Ching, L. L. (2004) *Organic Outperforms Conventional in Climatic Extremes*, Institute for Science in Society. www.i-sis.org.uk/OrganicOutperforms.php accessed in March 2008

Cochrane, W. W. (1958) *Farm Prices: Myth and Reality*, Minneapolis, University of Minnesota Press

Colosio, C. (1997) 'Dithiocarbamates and the immune system', *Food and Chemical Toxicology*, vol 35, no 3, p429

Conford, P. (2001) *The Origins of the Organic Movement*, Edinburgh, Floris Books

Conway, G. (1998) 'The doubly Green Revolution: a context for farming systems research and extension in the 21st century', in Stilwell, T. (ed) *Journal for Farming Systems Research-Extension. Special Issue 2001*, Tuscon, IFSA, pp1–16

Cook, D. (2004) *The Natural Step: Towards a Sustainable Society*, Schumacher Briefing 11, Devon, Green Books

Crucefix, D. (1998) *Organic Agriculture and Sustainable Rural Livelihoods in Developing Countries*, NRET working paper, Natural Resources Institute, Chatham

Darley, J., Room, D. and Rich C. (2006) *Relocalise Now! Getting Ready for Climate Change and the End of Peak Oil*, Post Carbon Institute, Canada, BC, New Society Publishers

Davis, D., Epp, M. and Riordan, H. (2005) 'Changes in USDA food composition data for 43 garden crops, 1950 to 1999', *Journal of the American College of Nutrition*, vol 23, no 66, pp669–682

Deffeyes, K. S. (2006) *Beyond Oil: the View From Hubbert's Peak*, New York, Hill & Wang

Delate, K. and Cambardella, C. A. (2004) 'Organic production. Agroecosystem performance during transition to certified organic grain production', *Agronomy Journal*, vol 96, pp1288–1298

Desai, P. and Riddlestone, S. (2002) *Bioregional Solutions for Living on One Planet*, Schumacher Briefing No. 8, Devon, Green Books

Dewailly, E., Ayotte, A., Bruneau, S., Gingras, S. and Roy, R. (2000) 'Susceptibility to infections and immune status in Inuit infants exposed to organochlorines', *Experimental Health Perspectives*, vol 108, no 3, pp205–211

Diver, S. (2000) *Nutritional Quality of Organically Grown Food*, Arkansas, Appropriate Technology Transfer for Rural Areas (ATTRA)

Dragsted, L., Loft, S., Basu, S., Jakobsen, J., Hansen, J., Pederson, A., Sandstrøm, B., Kall, M., Rasmussen, S. E., Hermetter, A., Haren, G. R., Breinholt, V., Castenmiller, J. J. M., Stagsted, J. and Skibsted, L. (2004) 'The 6-a-day study: effects of fruit and vegetables on oxidative stress and antioxidant defence in healthy non-smokers', *American Journal of Clinical Nutrition*, vol 79, no 6, pp1060–1072

Ellis, K. A., Innocent, G., Grove-White, D., Cripps, P., McLean, W. G., Howard, C. V. and Mihm, M. (2006) 'Comparing the fatty acid composition of organic and conventional milk', *Journal of Dairy Science*, 89, pp1938–1950

Fairlie, S. (2007) 'Can Britain feed itself?' *The Land* (Winter), pp18–26

FAO (1997) *The State of the World's Plant Genetic Resources for Food and Agriculture*, Rome, FAO

FAO (2002) *Food Needs – How Much is Enough? Revising Human Energy Requirements.* News & Highlights (17 January) www.fao.org/NEWS/2002/ 020103-e.htm

FAO (2006) *The State of Food Security in the World, 2006*, Rome, FAO

Farm Africa (2004) *Reaching the Poor: A Call to Action. Investment in Smallholder Agriculture in Sub-Saharan Africa*, London, Farm Africa/Harvest Help/Imperial College

Faustini, A. (1996) 'Immunosuppressive effects of Chlorophenoxy herbicides', *Food and Chemical Toxicology*, no 34, pp1190–1191

Feenstra, G. (2002) 'Creating space for sustainable food systems: lessons from the field', *Agriculture and Human Values*, vol 19, no 2, pp99–106

Fernándes, L., Mejía del Cid, J., Nishat, M., Wright, J. and Zabaleta, P. (1997) *Building a Bridge to So José do Norte: Diagnosis for Research Needs of Family Farms*, Wageningen, International Centre for Development-Oriented Research in Agriculture

Finamore, A., Britti, M. S., Roselli, M., Bellovino, D., Gaetani, S. and Mengheri, E. (2004) 'Novel approach for food safety evaluation. Results of a pilot experiment to evaluate organic and conventional foods', *Journal of Agricultural and Food Chemistry*, 52, pp7425–7431

Friedman, M. (2006) 'Potato glycoalkaloids and metabolites: roles in the plant and in the diet', *Journal of Agricultural and Food Chemistry*, 54, pp8655–8681

Goewie, E. A. (2002a) 'Organic agriculture in the Netherlands: developments and challenges', *Wageningen Journal of Life Sciences*, vol 50, no 2, pp153–170

Goewie, E. A. (2002b) 'Organic production. What is it?' *Urban Agriculture Magazine*, no 6 (April), pp5–8

Granstedt, A. G. and Kjellenberg, I. (1997) 'Long term field experiment in Sweden: effects of organic and inorganic fertilisers on soil fertility and crop quality', in Lockeretz, W. (ed) *Agricultural Production and Nutrition, Proceedings of an International Conference* (19–21 March), School of Nutrition Science and Policy, Tufts University, Boston MA, pp79–90

Greene, C. and Kremen, A. (2003) *US Organic Farming in 2000–2001: Adoption of Certified Systems*, Washington DC, US Department of Agriculture, Economic Research Service, Resource Economics Division, Agriculture Bulletin No. 780

Hall, A. and Mogyorody, V. (2001) 'Organic farmers in Ontario: an examination of the conventionalisation argument', *Sociologia Ruralis*, vol 41, no 4 (October), pp400–422

Hamer, E. and Anslow, M. (2008) 'Ten reasons why organic farming can feed the world', *The Ecologist*, vol 38, issue 2 (March), pp 43–46

Hancock, D. B., Martin, E. R., Mayhew, G. M., Stajich, J. M., Jewett, R., Stacy, M. A., Scott, B. L., Vance, J. M. and Scott, W. K. (2008) 'Pesticide exposure and risk of Parkinson's disease: a family-based case-control study', *BMC Neurology*, vol 8, no 6, p12

Harnapp, V. (1988) Food self-sufficiency or food dependency, *International Conference on Sustainable Agriculture* (September), Columbus OH

Harris, P. (1998) 'Constraints to conversion in Sub-Saharan Africa', *Ecology and Farming*, no 18, Germany, IFOAM

Hawken, P. (1993) *The Ecology of Commerce*, New York, Harper Collins

Hazell, P. (1995) *Managing Agricultural Intensification*, 2020 Brief 11 (February), Washington DC, IFPRI

Hazell, P. and Garrett, J. L. (2001) *Reducing Poverty and Protecting the Environment: the Overlooked Potential of Less-favored Lands*, 2020 Brief 39 (June), Washington DC, IFPRI

Heaton, S. (2001) *Organic Farming, Food Quality and Human Health: A Review of the Evidence*, Bristol, Soil Association

Heinäaho, M., Pusenius, J. and Julkunen-Tiitto, R. (2006) 'Organic farming methods affect the concentration of phenolic compounds in sea buckthorn leaves'. Paper presented at *Joint Organic Congress*, Odense, Denmark (30–31 May)

Heinberg, R. (2007) *The Party's Over: Oil, War and the Fate of Industrial Societies*, London, Clairview Books

Heller, M. and Keoleian, G. (2000) *Life-Cycle Based Sustainability Indicators for Assessment of the US Food System*, Ann Arbor MI, Center for Sustainable Systems, University of Michigan

Hirsch, R. L., Bezdek, R. and Wendling, R. (2005) *Peaking of World Oil Production – Impacts, Mitigation and Risk Management*, National Energy Technology Laboratory, US Department of Energy. http://www.netl.doe.gov/publications/others/pdf/Oil_Peaking_NETL.pdf accessed May 2008

Ho, M.-W. and Ching, L. L. (2008) 'Mitigating climate change through organic agriculture and localised food systems', *Science in Society*, issue 37 (Spring), pp47–51

Hole, A. G., Perkins, A. J., Wilson, J. D., Alexander, I. H., Grice, P. V. and Evans, A. D. (2005) 'Does organic farming benefit biodiversity?' *Biological Conservation*, vol 122, pp113–130

Hood, C. (1998) *The Art of the State. Culture, Rhetoric and Public Management*, Oxford, Clarendon Press

Holt-Giménez, E. (2002) 'Measuring farmers' agroecological resistance after Hurricane Mitch in Nicaragua: a case study in participatory, sustainable land management impact monitoring', *Agriculture, Ecosystems and Environment*, no 93, pp87–105

Hopkins, R. (2006) *Energy Descent Pathways: Evaluating Potential Responses to Peak Oil*, MSc dissertation, University of Plymouth. www.transitionculture.org

Hopkins, R. (2007) 'The transition towns concept', Presentation at *Soil Association Annual Conference: One Planet Agriculture: Preparing for a post-peak oil food and farming future* (26 January), Bristol, The Soil Association

Howard, A. (1940) *An Agricultural Testament*, Oxford, Oxford University Press

Huang, R. Q. (1998) *Coping Behaviour of Extension Agents in Role Conflict Situations*, doctoral thesis, Wageningen University

Hubbert, M. K. (1956) 'Nuclear energy and the fossil fuels', *Drilling and Production Practice, 1956*, American Petroleum Institute. http://www.energybulletin.net/node/13630 accessed April 2006

Hyems, E. (1952) *Soil and Civilisation*, London, Thames & Hudson

IAC (2003) *IAC Study on Science and Technology Strategies for Improving Agricultural Productivity and Food Security in Africa*, progress report (May) (unpublished), Interacademy Council Study Panel

IFOAM (1998) *Basic Standards for Organic Production and Processing*, Germany, IFOAM

IFOAM (2008a) http://www.ifoam.org/about_ifoam/principles/index.html accessed in March 2008

IFOAM (2008b) *Organic Agriculture and HIV/AIDS in Sub-Saharan Africa*, Germany, IFOAM

Johns, T., Smith, I. F. and Eyzaguirre, P. B. (2006) 'Agrobiodiversity, nutrition and health', in Hawkes, C. and Ruel, M. T. (eds) *Understanding the Links Between Agriculture and Health*, Washington DC, IFPRI, p12

Kaltoft, P. (2001) 'Organic farming in late modernity: at the frontier of modernity or opposing modernity?' *Sociologia Ruralis*, vol 41, no 1 (January), pp146–158

Kindell, H. H. and Pimentel, D. (1994) 'Constraints to the expansion of global food supply', *Ambio*, vol 23, no 3 (May)

Kloppenburg, J. (1991) 'Social theory and the de/reconstruction of agricultural science: local knowledge for alternative agriculture', *Rural Sociology*, vol 56, no 4, pp519–548

Kloppenburg, J., Hendrickson, J. and Stevenson, G. W. (1996) 'Coming into the foodshed', *Agriculture and Human Values*, vol 13, no 3 (Summer), pp33–42

Kolb, D. (1984) *Experiential Learning. Experience as a Source of Learning and Development*, Englewood Cliffs NJ, Prentice Hall

Kumar, V., Mills, D., Anderson, J. and Mattoo, A. K. (2004) 'An alternative agriculture system is defined by a distinct expression profile of select gene transcripts and proteins', *Proceedings of the Academy of Sciences*, vol 101 (July), pp10535–10540

Lampkin, N. (1990) *Organic Farming*, Ipswich, Farming Press

Lang, T. (1997) 'The challenge for food policy', Schumacher Lecture. *Resurgence*, vol 181 (March–April)

Lang, T. (2008) 'Food insecurity', *The Ecologist*, vol 38, issue 2 (March), pp32–34

Lang, T. and Heasman, M. (2004) *Food Wars: The Global Battle for Mouths, Minds and Markets*, London, Earthscan Publications

Lauridsen, C., Jørgensen, H., Halekoh, U., Christensen, L. P. and Brandt, K. (2005) 'Organic diet enhanced the health of rats', *DARCOF E-news*, newsletter from Danish Research Centre for Organic Farming (March). http://www.darcof.dk/cnews/mar05/health.html

Leiber, F., Huber, F., Dlugosch, G. E. and Fuchs, N. (2006) 'How to evaluate the influences of organic food on the whole human being? Experiences with a convent study', presentation at *6th Europäische Sommerakademie für Biolandwirtschaft*, Czech Republic, 29 June–1 July

Lotter, D. W., Seidel, R. and Liebhart, W. (2003) 'The performance of organic and conventional cropping systems in an extreme climate year', *American Journal of Alternative Agriculture*, vol 18, no 3, pp146–154

Lu, C., Toepel, K., Irish, R., Fenske, R., Barr, D. B. and Bravo, R. (2005) 'Organic diets significantly lower children's dietary exposure to organophosphorous pesticides in food', *Environmental Health Perspectives*, vol 114, pp260–263

Lucas, C., Jones, A. and Hines, C. (2006) *Fuelling a Food Crisis: the Impact of Peak Oil on Food Security*, The Greens/European Free Alliance. http://www.energybulletin.net/node/24319 accessed May 2008

MacRae, R. J., Hill, S. B., Henning, J. and Bentley, A. J. (1990) 'Policies, programs and regulations to support the transition to sustainable agriculture in Canada', *American Journal of Alternative Agriculture*, vol 5, no 2, pp76–92

MacRae, R. J., Henning, J. and Hill, S. B. (1993) 'Strategies to overcome barriers to the development of sustainable agriculture in Canada: the role of agribusiness', *Journal of Agricultural and Environmental Ethics*, vol 6, no 1, pp21–51

Mäder, P., Fliessbach, A., Dubois, D., Gunst, L., Fried, P. and Niggli, U. (2002) 'Soil fertility and biodiversity in organic farming', *Science*, vol 226, pp1694–1697

Magkos, F., Arvanti, F. and Zampelas, A. (2003) 'Organic food: nutritious food or food for thought? A review of the evidence', *International Journal of Food Sciences and Nutrition*, no 54, pp357–371

Mansvelt, J. D. van and Lubbe, M. J. van der (1999) *Checklist for Sustainable Landscape Management*. Final report of the EU concerted action AIR-CT93–1210: The landscape and nature production capacity of organic/sustainable types of agriculture, Amsterdam, Elsevier publications

Marriot, E. E. and Wander, M. M. (2006) 'Total and labile soil organic matter in organic and conventional farming systems', *Soil Science Society of America Journal*, no 70, pp950–959

Massingham, H. J. (1942) *Remembrance*. London, Batsford Press

Mayer, A. M. (1997) 'Historical changes in the mineral content of fruits and vegetables', *British Food Journal*, vol 99, no 6, p207

Mazoyer, M. and Roudart, L. (2006) *A History of World Agriculture, from the Neolithic Age To The Current Crisis*, London, Earthscan Publications

McCance, R. A. and Widdowson, E. M. (1940–1991) *The Composition of Foods*, 1st to 5th editions, London, MAFF/Royal Society of Chemistry

McCarrison, R. A. (1945) *Farming and Gardening for Health or Disease*, London, Faber

McKillop, A. (2005) *The Final Energy Crisis*, London, Pluto Press

McNeely, J. A. and Scherr, S. (2001) *Common Ground, Common Future: How Ecoagriculture Can Help Feed the World and Save Wild Biodiversity*, Report No. 5/01, Gland, IUCN

Merlo, S. (1997) Successful sustainable development with government support. *Conference Proceedings: The Future for Organic Trade*, 5th IFOAM International Conference on Trade in Organic Products, 24–27 September, Germany, IFOAM

Mesure, S. (2008) 'The £20 bn food mountain: Britons throw away half of the food produced each year', *The Independent* (Sunday 2 March)

Mitchell, A. E., Hong, Y. J., Koh, E., Barrett, D. M., Bryant, D. E., Denison, R. F. and Kaffka, S. (2007) 'Ten year comparison of the influences of organic and conventional crop management practices on the content of flavonoids in tomatoes', *Journal of Food and Agricultural Chemistry*, vol 55, no 15, pp6154–6159

Mobbs, P. (2005) *Energy Beyond Oil*, Leicester, Matador Books

Mollison, B. (1988) *Permaculture, a Designer's Manual*, Australia, Tagari Publications

Moreira, M., Roura, S. I. and del Valle, C. E. (2003) 'Quality of Swiss chard produced by conventional and organic methods', *Food Science and Technology*, vol 36, pp135–141

NAS (1989) *Alternative Agriculture*, Washington DC, National Academic Press

Niggli, U., Earley, J. and Ogorzalek, K. (2007) 'Organic agriculture and environmental stability of the food supply', Issues Paper: Stability of Food Supply. *International Conference on Organic Agriculture and Food Security*, 3–5 May, Rome, FAO

Niles, J., Brown, S., Pretty, J., Ball, A. and Fay, J. (2001) *Potential Carbon Mitigation and Income in Developing Countries from Changes in Use and Management of Agricultural and Forest Lands*, Occasional Paper 2001–04, University of Essex, Centre for Environment and Society

Norberg-Hodge, H. (2002) *Bringing the Local Economy Home – Local Alternatives to Global Agribusiness*, London, Zed Books

Offerman, F. and Nieberg, H. (2000) *Economic Performance of Organic Farms in Europe*. Organic Farming in Europe: Economics and Policy Volume 5, Germany, University of Hohenheim

Oldeman, L. R. (1999) 'Soil degradation: a threat to food security?' in Bindraban, P. S., Keulen, H. van, Kuyvenhoven, A., Rabbinge, R. and Uithol, P. W. J. (eds) *Food Security at Different Scales: Demographic, Biophysical and Socio-economic Considerations*, Quantitative Approaches in Systems Analysis No. 21, Wageningen University, pp105–117

Oram, J. (2008) 'Local solution to global food crisis', leaders and reply, *The Guardian* (Wednesday 27 February), p33

Parrott, N. (2002) 'The real green revolution', *Ecology and Farming*, vol 30 (May–August), Germany, IFOAM, pp5–7

Parrott, N. and Marsden, T. (2002) *The Real Green Revolution. Organic and Agroecological Farming in the South*, London, Greenpeace Environmental Trust

Pedersen, H. L. and Bertelsen, M. (2002) 'Alleyway groundcover management and scab resistant apple varieties', *ECO-FRU-VIT. 10th International Conference on Cultivation Techniques and Phytopathological Problems in Organic Fruit Growing and Viticulture*, pp19–21. www.infodienst-mlr.bwl.de/la/lvwo/ecofruvit/alleywayground3.pdf

Pfeiffer, D. A. (2006) *Eating Fossil Fuels. Oil, Food and the Coming Crisis in Agriculture*, Canada, New Society Publishers

Picton, L. P. (1943) *Thoughts on Feeding*, London, Faber & Faber

Pimentel, D., Hepperly, P., Hanson, J., Douds, D. and Seidel, R. (2005) 'Environmental, energetic and economic comparisons of organic and conventional farming systems', *Bioscience*, vol 55, no 7, pp573–582

Pingali, P. L. and Rosegrant, M. W. (1998) 'Supplying wheat for Asia's increasingly westernised diets', *American Journal of Agriculture*, vol 80, no 5, pp954–959

Plaskett, L. G. (1999) 'Nutritional therapy to the aid of cancer patients', submitted to *International Journal of Alternative and Complementary Medicine* (December). http://www.therapyofcancer.co.uk/articles.cfm accessed June 2008

Pretty, J. N. (1995) *Regenerating Agriculture. Policies and Practice for Sustainability and Self-Reliance*, London, Earthscan Publications

Pretty, J. (1998) *The Living Land*, London, Earthscan Publications

Pretty, J. (2002) *Agri-Culture. Reconnecting People, Land and Culture*, London, Earthscan Publications

Pretty, J. and Shaxson, F. (1997) *The Potential of Sustainable Agriculture*, paper presented at the DFID Natural Resources Advisers Conferences, London, DFID

Pretty, J., Thompson, J. and Hinchcliffe, F. (1996) *Sustainable Agriculture: Impacts on Food Production and Food Security*, Gatekeeper Series, 60, London, IIED

Pretty, J. N., Brett, C., Gee, D., Hine, R. E., Mason, C. F., Morison, J. I. L., Raven, H., Rayment, M. and van de Bijl, G. (2000) 'An assessment of the external costs of UK agriculture', *Agricultural Systems*, vol 65, no 2, pp113–136

Pretty, J. N., Morrison, J. I. L. and Hines, R. E. (2002) 'Reducing food poverty by increasing agricultural sustainability in developing countries', *Agriculture, Ecosystems and the Environment*, vol 95, pp217–234

Raupp, J. (1996) 'Quality investigations with products of the long term fertilisation trial in Darmstadt – second period: fertilisation with total nitrogen equivalents', *Quality of Plant Products Grown with Manure Fertilisation: Proceedings of the Fourth Meeting*, Juva, Finland, 6–9 July, Darmstadt, Institute for Biodynamic Research, pp13–33

Rees, W. and Wackernagel, M. (1996) 'Urban ecological footprints: why cities cannot be sustainable', *Environmental Impact Assessment Review*, no 16, pp223–248

Rembialkowska, E. (2005) 'Quality of plant products from organic agriculture', Technical Seminar 11, *Proceedings, First Annual Congress of the European Union Project QualityLowInputFood*, 6–9 January, University of Newcastle

Rembialkowska, E., Hollman, E. and Rusakzonek, A. (2007) 'Influencing a process on bioactive substances content and antioxidant properties of apple puree from organic and conventional production in Poland', *Improving Sustainability in Organic and Low Input Food Production Systems; Proceedings of the 3rd International Congress of European Integrated Project Quality Low Input Food*, March, University of Hohenheim, Germany

van Rensburg, W. J., Venter, S. L., Netshiluvhi, T. R., van den Heever, E., Vorster, H. J., Repetto, R. and Balinga, S. S. (2004) *Pesticides and the Immune System*, Washington DC, World Resources Institute

Röling, N. (2002) 'Beyond the aggregation of individual preferences. moving from multiple to distributed cognition in resource dilemmas', in Leeuwis, C. and Pyburn, R.

(eds) *Wheelbarrows Full of Frogs. Social Learning in Natural Resource Management*, Assen, Koninklijke Van Gorcum, pp25–28

Röling, N. G. and Wagemakers, M. A. E. (eds) (1998) *Facilitating Sustainable Agriculture*, Cambridge, Cambridge University Press

Rotmans, J., Kemp, R. and van Asselt, M. (2001) 'More evolution than revolution: Transition Management in policy', *Foresight. The Journal of Future Studies, Strategic Thinking and Policy*, vol 3, no 1, pp1–17

Rundgren, G. (2002) *Organic Agriculture and Food Security*, Dossier 1, Germany, IFOAM

SAN (2003) *Opportunities in Agriculture: Transitioning to Organic Production*, Sustainable Agriculture Network, USDA-SARE, Washington DC

Sattler, F. and Wistinghausen, V. (1992) *Biodynamic Farming Practice*, Cambridge, Cambridge University Press

Scialabba, N. E. and Hattam, C. (2002) *Organic Agriculture, Environment and Food Security*, Environment and Natural Resources Series 4, Rome, FAO

Schuman, M. (2000) *Going Local: Creating Self-reliant Communities in a Global Age*, New York, Routledge

Simmons, M. R. (2005) *Twilight in the Desert: The Coming Saudi Oil Shock and the World Economy*, Hoboken NJ, John Wiley & Sons

Sinha, R. K. (1997) 'Embarking on the second green revolution for sustainable agriculture in India: a judicious mix of traditional wisdom and modern knowledge in ecological farming', *Journal of Agricultural & Environmental Ethics*, vol 10, no 2, pp183–197

Smith, P. (1989) *Management in Agricultural and Rural Development*, Elsevier Applied Science, London

Soil Association (2007) *The Nutritional Benefits of Organic Milk – A Review of the Evidence*, Bristol, Soil Association

Souza, J. L. (1998) *Agricultura organica – Tecnologias para a producào organica de alimentos saudáveis*, vol 1, Spain, EMCAPA, Domingos Martins

Stamp, N. (2003) 'Out of the quagmire of plant defence hypotheses', *The Quarterly Review of Biology*, vol 78, pp23–55

Stapledon, R. G. (1942) *The Land Now and Tomorrow*, London, Faber & Faber

Strahan, D. (2007) *The Last Oil Shock*, London, McArthur & Co./ John Murray

Sustain (2003) *Myth and Reality, Organic versus Non-Organic*, London, Sustain

Tansey, G. and Worsley, T. (1995) *The Food System*, London, Earthscan Publications

Tertzakian, P. (2007) *A Thousand Barrels a Second: The Coming Oil Break Point and the Challenges Facing an Energy Dependent World*, McGraw-Hill Professional, New York

Toor, R. K., Savage, G. P. and Heeb, A. (2007) 'Influence of different types of fertilisers on the major antioxidant components of tomatoes', *Journal of Food Composition and Analysis*, vol 19, pp20–27

Transition Towns (2008) http://transitiontowns.org/TransitionNetwork/TransitionCommunities accessed in March 2008

Vidal, J. (2003) 'Farmers are dying out', *The Guardian* (7 August). http://www.guardian.co.uk/uk/2003/aug/07/ruralaffairs.johnvidal

Voccia, I., Blakley, B., Brousseau, P. and Fournier, M. (1999) 'Immunotoxicity of pesticides: a review', *Toxicology and Industrial Health*, vol 15, no 102, pp119–32

Vorley, B. (2003) *Food, Inc. Corporate Concentration from Farmer to Consumer*, UK Food Group. www.agribusinessaccountability.org/page/243/

WHO (1999) *Public Health Impact of Pesticides Used in Agriculture*, Geneva, WHO/UNEP

Winter, C. K. and Davis, S. F. (2006) 'Organic foods', *Journal of Food Science*, vol 71, no 9, pp117–124

Wolfert, J. (2002) *Sustainable Agriculture: How to Make it Work? A Modelling Approach to Support Management of a Mixed Ecological Farm*, published PhD thesis, Wageningen University

Woodward, L. (1999) 'Organic farming, food quality and human health', *Proceedings, 11th National Conference on Organic Food and Farming*, 8–10 January, Royal Agricultural College, Cirencester

Woodward, L. (2001) 'The scientific basis of organic food and farming', *Proceedings, British Association for the Advancement of Science*, 7 September, Glasgow, www.efrc.com/efrc/scientificbasis.htm

Worthington, V. (2001) 'Comparison of nutrient levels in organic food compared to non-organic', *Journal of Alternative and Complementary Medicine*, vol 7, no 2, pp161–173

Wrench, G. T. (1972) *The Wheel of Health*, USA, Schocken Books

Wright, J. (2004) 'Going with the grain: mainstreaming an ecological paradigm for food security', paper presented at: *LEAD International Workshop on Sustainable Agriculture and the Global Food Economy*, April, Wye College. www.leadinternational.org.uk

Wright, J. (2005) *Falta Petroleo. Perspectives on the Emergence of a More Ecological Farming and Food System in Post-Crisis Cuba*, doctoral thesis, Wageningen University

WUR (2002) *Wageningse Visies op Voedseizekerheid. Wageningen Platform for Food Security*, Wageningen, Wageningen University & Research Centre

Ziesemer, J. (2008) *Energy Use in Organic Food Systems*, Rome, FAO

3

Researching Cuba

The enigma of Cuba

Cuba's fuel, input and import crisis may provide a unique example of an entire nation that has been forced to adapt to a situation of scarcity of petroleum and petroleum-based products; one that, it is assumed, had to increase its reliance on local and organic strategies. In the absence of any broad and systematic analysis during the mid- to late-1990s, including from Cuban-based organizations, the anecdotal reporting described in Chapter 1 is all there was to go on. Had Cuba been able to feed itself without petroleum inputs, and, whether or not this was so, what were policies and strategies that had been implemented in a bid to achieve food security? Was Cuban agriculture organic? Even if dramatic changes had not evolved, what were the coping and learning strategies emerging in this sudden and enforced situation? Would Cuba choose to remain this way even after its 'Special Period'?

Research perspectives

The material for this book is based on doctoral research that set out to answer these questions. Conducted between 1998 and 2005, this research was based at Wageningen University in the Netherlands, and in Cuba at the National Institute of Agricultural Sciences, INCA (Instituto Nacional de Ciencias Agrícolas). The research objective was to evaluate the implications for the food and agriculture sectors in Cuba of a widespread reduction in petroleum-based inputs. As discussed in Chapter 2, the issues being examined were complex and multidisciplinary, and were categorized into three perspectives for analysis: first, a perspective with which to determine the kind of low-petroleum food system that Cuba had developed; second, a perspective that addressed the issue of how Cuba managed the transition and change; and third, a perspective through which to ascertain the degree to which Cuba had managed to achieve an adequate food system for its people.

The study examined the interrelationships and driving forces of the different factors and actors involved in the food system, based as far as possible on first-hand evidence and on data available from within the country. A framework was designed with which to capture the external impacts resulting from the crisis, the individual and collective perceptions and coping strategies, and the mechanisms of the institutional support system. This framework was based on the rural livelihoods concept, which purposefully seeks to capture the dynamics of rural production and identify adaptive and coping strategies in times of crisis (Scoones, 1998; Khanya, 2000; DFID, 2001). Some adaptation of the standard livelihoods model was made, particularly in recognition of the central role of the state, and that markets were not necessarily a strong influencing force in the country. The framework is shown in Figure 3.1.

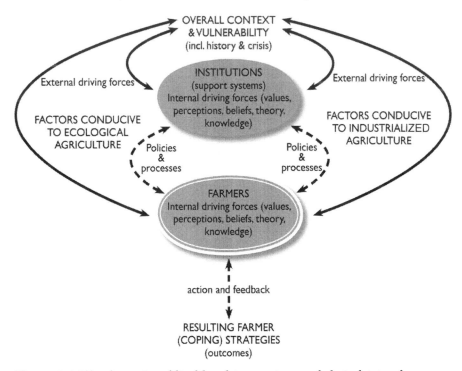

Figure 3.1 *The dynamics of livelihood interactions and their driving forces as a theoretical framework for research in Cuba*

Source: Wright, 2005

Places and people in the field

Field research was undertaken in Cuba between 1999 and 2001, under the auspices of the National Institute of Agricultural Sciences (INCA) which pertains to the Ministry of Higher Education (MES). The main study regions were the Provinces of Havana (municipalities of San Antonio de los Baños and Batabanó), Cienfuegos towards the centre of the island (municipality of

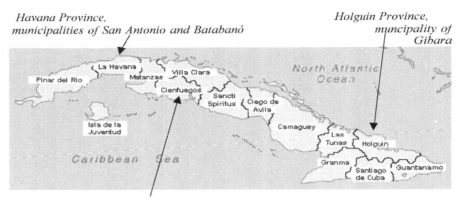

Havana Province,
municipalities of San Antonio and Batabanó

Holguin Province,
municipality of
Gibara

Cienfuegos Province
municipality of Cumanayagua

Figure 3.2 *Map showing the main study areas across three provinces of Cuba*
Source: Wright, 2005

Cumanayagua) and Holguin to the east (municipality of Gibara). In all three provinces, comparisons were made between the more independent Credit and Service Cooperatives (CCS) and the former state-controlled Basic Units of Cooperative Production (UBPC). Figure 3.2 shows these study areas. Briefer studies were also made in the provinces of Pinar del Rio, Villa Clara and Las Tunas, and a nationwide survey also included the provinces of Guantanemo, Ciego de Avila, Granma and Santiago de Cuba.

Formal and informal information gathering methods were used with a total of 414 Cubans comprising farmers and farm workers, farm management and administrative staff, extension, research and managerial staff of the Ministries of Agriculture, Higher Education, Public Health and Science, Technology and the Environment, and households, in largely rural but also urban areas. Box 3.1 lists the institutions that participated, either at national, provincial or municipal level.

Libraries and documentation centres were also visited. In addition, information was gathered from various scientific and development conferences, workshops and study tours held in Cuba during that period, and from practical farm work. Following the ethics of social science research, quotations from the field are unattributed, and no sources of primary information are given in this book.

Red threads affecting the research process

Any attempt to evaluate the production and food system of another country is ambitious. In order to simplify the line of investigation, the research focused on

Box 3.1 *Main institutions visited for field data collection, at national, provincial and municipal levels, in Cuba during 1999–2001*

Ministry of Agriculture and associated entities:
- Departments of Urban Agriculture (including Agricultural Services Shops), Science and Technology, Quality Control
- National Institutes of Soils, Plant Health (INISAV)
- provincial subdelegations
- Enterprises of Mixed Crops, Livestock, Citrus, Seeds, Select Fruits
- Seed Inspection and Certification Service
- Centres for the Reproduction of Entomophagens and Entomopathogens (CREE)
- Research institutes: Soils Research Institute, Plant Health Research Institute (Sanidad Vegetal), National Research Institute of Tropical Agriculture (INIFAT), National Research Institute of Tropical Roots and Tubers (INIVIT), Horticultural Research Institute 'Liliana Dimitrova' (IIHLD), Pastures and Forages Research Institute (IIPF), Institute of Agricultural Mechanisation (IIMA), Irrigation and Drainage Research Institute (IIID), Territorial Agricultural Research Station of Holguin (ETIAH)
- Acopio
- National Association of Small Producers (ANAP)
- Cuban Association of Agricultural and Forestry Technicians (ACTAF)
- Cuban Association of Organic Agriculture (ACAO).

Ministry of Higher Education:
- National Agricultural University of Havana (UNAH) (including the Centre for Studies in Sustainable Agriculture (CEAS))
- National Institute of Agricultural Sciences (INCA)
- Institute for Animal Sciences (ICA)
- National University of Havana (UNH) (Faculties of Geography and Sociology)
- Central University of Las Villas (UCLV)
- University of Holguin, University of Cienfuegos.

Ministry of Science, Technology and Environment (CITMA):
- Research Institute of Ecology and Systematics (IIES)
- Tropical Geography Institute (IGT)
- Centre for Information, Management and Environmental Education (CIGEA)
- Foundation for Nature and Humanity Antonio Nùñez Jimenez (FANJ)
- Botanic Gardens of Cienfuegos
- Institute of Plant Biotechnology, Genetics and Biotechnology Research Centre (CIGB).

Ministry of Food Industry:
- Institute of Nutrition and Food Hygiene (INHA)

National Statistics Office (ONE).

Regional offices of:
- Food and Agriculture Organization (FAO)
- European Union (EU).

Source: Wright, 2005

non-state farms which at that time were working 70 per cent of cultivable land. This focus also made it easier to obtain research permission. Additionally, the research avoided the main export crops of sugarcane and tobacco, because these crops were receiving priority treatment for inputs, which was not the norm. The degree of use of agrochemicals was chosen as a 'red thread', around which other factors could be woven in. This choice was based on the assumption that the loss of agrochemical inputs would be one of the main driving forces for change in the food and farming system. Early on in the process, however, it became clear that agrochemicals were considered of lesser importance; almost all farmers and others interviewed agreed that the main impact, and main challenge to change, was *Falta petroleo!* (Lack of oil!).

This lack of petrol affected the research process itself: for example, there was little transport to reach the research sites. The low level of resources also meant that books, journals and reports had not been published for several years and paper was scarce, so there was very little documentation on, for example, farm economic analyses, published research results, institutional annual reports, or land survey information.

Cuba's recent history and current socio-political situation also affected the research process. Because of the sanctions and other actions of the United States against Cuba, the country was understandably cautious of foreign researchers working on issues of potential domestic importance. This meant that it was easier to access grassroots and lower-level authorities than central government. During the period of this work, many other prospective foreign researchers came and went empty handed. Every country has its own particular regulations and conditions regarding international collaboration, and Cuba was no different. It was crucial to obtain official authorization through a Cuban institution, signed off by the relevant Ministry. This process of obtaining authorization might take 3–12 months to come through. It was also crucial to identify a topic of mutual interest, preferably one that complemented or supported existing work in the country, and one that was not dealing with sensitive information. Planning would need to take into account the low level of institutional resources available, especially, but not only, financial. In retrospect, probably the wisest approach for establishing collaboration would have been to make a short reconnaissance visit on a tourist visa, as part of an organized tour or conference delegation, in order to make direct visual contact and establish trust. With international interest in Cuba likely to remain at a consistent high, and with Cuba's increasing interest in strengthening international linkages, the mechanisms for organizing research will doubtless advance.

Another factor affecting the research was the unavoidably constructivist and interpretive nature of the work. Many divergent stories and perspectives emerged during the research interviews and discussions, and these were then subject to translation and interpretation by the researcher, who by being a foreigner influenced the type of information being supplied. It became clear that the more synthesizing, summarizing and construction of neat macro-views and generalizations was made, the more understanding and meaning would be lost, and especially when dealing with local complexity. There were already sufficient broad brush descriptions of Cuban experiences. Therefore, the information gathered was treated as a narrative collection of perspectives, and as much detail was included as relevance allowed. At the time this research was conducted, social-science studies were uncommon in the technology-focused agricultural sector. Agricultural libraries had filed under this heading books on political science. A sociologist commented that 'Little attention has been paid to the social aspects of production, the state doesn't think this to be necessary in a socialist country'. Outside the arts and humanities, social studies conducted in Cuba were frequently dealing with matters of national importance and on sensitive issues. Since that time, domestic awareness has increased on the significance of what the country has gone through, and there is also more flexibility within society to constructively analyse and evaluate. In view of this, first-hand accounts and more detailed works on the coping strategies of Cuba in dealing with the situation of the Special Period, written by Cubans themselves, are eagerly anticipated.

References

DFID (2001) *Sustainable Livelihoods Guidance Sheets*, London, Department for International Development

Khanya (2000) *Guidelines for Undertaking a Regional/National Sustainable Rural Livelihoods Study*, Bloemfontein, Khanya – managing rural change. www.khanya-mrc.co.za

Scoones, I. (1998) *Sustainable Rural Livelihoods: A Framework for Analysis*, Working Paper 72 (June) Sussex, IDS

Wright, J. (2005) *Falta Petroleo. Perspectives on the Emergence of a More Ecological Farming and Food System in Post-Crisis Cuba*, doctoral thesis, Wageningen University

4

The Historical Context: Cuban Agriculture and Food Systems of the 20th Century

The Socialist Revolution in Cuba was heavily backed by the rural population, and for over 40 years of its existence has continually emphasized the improvements in the farming and food system compared with the pre-Revolutionary period. Yet, as with food systems in most countries, it was highly vulnerable to changes in the external relations upon which it depended. A review of its historical context provides an explanation for this vulnerability and also describes the fundamental structures and mechanisms still visible in the farming and food system today. This review includes the events leading directly up to the crisis of the early 1990s and the short-term coping strategies that ensued.

The agro-geography of Cuba

Located just south of the Tropic of Cancer, the archipelago of Cuba, part of the Greater Antilles chain, is formed of approximately 1600 Caribbean islands, the largest of which – Cuba – lies only 160km south of the Florida Keys, and commands the main entry to the Gulf of Mexico. Long and narrow (described by Cubans as 'the sleeping alligator'), this island is 1200km in length, 40–290km in width, covers 107,467km^2, and is mountainous for 20 per cent of its land surface. Three different mountain ranges can be distinguished: the highest (up to 2000m) and most complex in the eastern provinces of Santiago de Cuba and Guantánamo; the central low ranges of Cienfuegos and Sancti Spiritus; and, to the west, the limestone hills of Pinar del Rio. The other 80 per cent of the land surface consists of more gentle rolling hills and extensive lowlands, with deep red and sandy clays, and fertile alluvial soils in the flood plains (Weeks and Ferbel, 1994; Hatchwell and Calder, 1995).

As the island is subtropical, relative humidity is high at 75 per cent, while annual temperature fluctuations are small, with an average minima and maxima of 22.6°C and 27.6°C. The rainy season lasts from May to October, and 80 per cent of total rainfall occurs during this time, followed by a dry season from November to April. Over the island, the average annual rainfall is just under

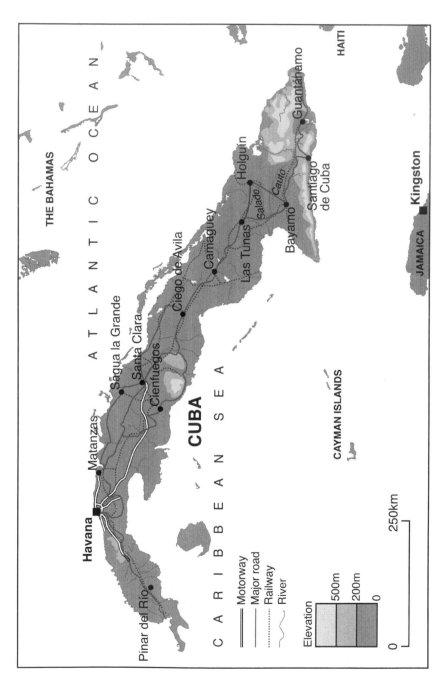

Figure 4.1 *Map of Cuba*

1500mm, with a minimum annual average of above 650mm (ONE, 2006). Regionally, rainfall is highest in the north-west of the island and mountain areas, whilst eastern regions are dry. Although rivers are plentiful, the island's narrow, elongated form means that much of this fresh water runs off quickly seawards, with little retention other than where captured by human intervention. The east coast is subject to hurricanes from August to October, and droughts are also common.

Just over half of the total land area, 6.7 million ha, is in agricultural use. Of this, 3.1 million ha is cultivated, almost 2 million ha under permanent crops – sugarcane, coffee, banana and citrus – and 1.2 million ha seasonally cropped with rice, horticulture and tobacco. Approximately 12 per cent of total land area is under irrigation (ODCI, 1998). The non-cultivated agricultural area consists of rough grazing and neglected land, much of which is covered by the thorny, invasive shrub *marabou* (*Dichrostachys cinerea*). Forests cover another 3 million ha of the country (ONE, 2006). Most recently available figures, from 2001, showed approximately one-third of the cultivable soils to be productive or very productive, and two-thirds of very low productivity, this being largely due to erosion, acidity, low organic matter and salinization (Duran, 1998; ONE, 2006). Agriculture and fisheries account for 3 per cent of GDP[1] in Cuba (ONE, 2006), although historically they have made a major contribution to the Cuban economy, with sugar, tobacco, coffee, alcohol and fish being important export products. Other major components of the economy are chemicals, construction services, nickel, steel, cement, agricultural machinery and biotechnology.

Sugarcane still covers a relatively large area throughout the country. It is grown on the better soils, and, because of the highly mechanized production approach, on flat areas. Tobacco, the most important export crop, is traditionally grown in the provinces of Pinar del Rio, Holguin and Sancti Spiritus. Potato, an important and popular staple of the national diet, is produced largely in the provinces of Havana and Matanzas, which have the most fertile and deep soils on the island. Large-scale horticulture is predominantly centred around the capital – in Havana and Matanzas Provinces – while grazing and permanent pastures are concentrated in the eastern provinces with their less fertile, more erodable and shallow soils. Tree crops such as coffee, cocoa and citrus are found in the mountainous eastern regions of Santiago de Cuba and Guantanemo.

Cuba has a population of 11,239,000, 75 per cent of whom reside in urban areas and 20 per cent in Havana city (ONE, 2006). Statistics on ethnic composition vary: Funes (2002) suggests that the population is largely Caucasian (66 per cent), with 22 per cent being mixed (*mestizo* or *mulatto*) and 12 per cent black, whereas ODCI (1998) identified 37 per cent as being Caucasian and 51 per cent mixed. Population growth rate stood at 0.2 in 2005, literacy at 95.7 per cent in 1998, and life expectancy at 77 years in 2001–03 (ODCI, 1998; ONE, 2006).

The food production system up to 1989

Pre-Columbian food production

Cuba's agricultural history has been one of colonialization. The original inhabitants of Cuba, the Guanahatabeyes, are believed to have arrived there about 10,000 years ago from North America and were primarily fisherfolk and gatherers. These were joined, 4500 years ago, by migrants from South America known as Ciboneyes. They maintained their traditional lifestyles until the arrival of the Taino tribe, an offshoot of the Arawaks from South America, 1500 years ago. The Taino were hunters but also settled farmers who cropped a large proportion of their diet, mainly maize and cassava supplemented with sweetpotato, squash, beans, peanuts and fruits such as guayaba, guanabana and pineapple. Cassava in particular played an important role: the bitter variety was grated, leeched and toasted on ceramic griddles to produce a bread known as *casabe*. This bread, a central part of the diet, could be stored for long periods of time without spoiling. Religious ceremonies focused around cassava production and the provider of cassava or *Yucahu*. The Tainos possessed a good knowledge of ecological aspects of agriculture. For example, they used assemblages of species with different growth patterns so as to minimize competition for soil and moisture resources – often on elevated mounds or *conucus* – and they also used nitrogen-fixing legumes to enhance soil fertility. They practised slash-and-burn agriculture, which, given the low population density, was relatively sustainable. They eventually fell victim to European colonizers, and little record was left of their cultural history, although remnants of their existence, and descendants, can still be found in Cuba (Rosset and Benjamin, 1994; Weeks and Ferbel, 1994).

Colonial food production

From a European perspective, the major group of colonialists was the Spanish, Columbus arriving and claiming the island for Spain in 1492. By 1511, land was being distributed to Spanish settlers who started to develop sugarcane plantations. In the 18th century, over 650,000 African slaves – mainly from Nigeria – were brought over to work these plantations. Coupled with the extensive and large-scale cattle ranches, a sharp contrast was already emerging with the small- and medium-scale crop-based systems and farmers. By the 19th century, the sugarcane plantations had become the most mechanized in the world, and were producing almost one-third of the world's sugar. Their growth was dependent on slave labour, but after the abolition of slavery in Cuba in 1886, manpower was augmented by the arrival of indigenous Mexicans and Chinese.

Towards the end of the 19th century, the United States increasingly became the major agricultural trading partner and financier. After the successful War of Independence against Spanish rule, from 1895 to 1898, a small group of US farmers came to dominate ownership of the main plantations: 13 of the sugar-cane *latifundios* were producing 70 per cent of total sugar output, and within 40 years the number of small, diversified farms fell by more than half, from 90,000 in 1895 to 38,130 by 1934. At this time, 95 per cent of land was in private hands, and Cuban small farmers held land – *minifundios* – largely through tenancy, sub-tenancy, share-holding and land administration. In contrast to the intensively farmed *latifundios*, these small farmers practised more traditional methods, with a low use of inputs. In 1945, for example, their irrigated land occupied only 1–2 per cent of total agricultural area, no pesticides were used, and these farmers did not generally own tractors or other machinery. Feeling exploited by the US monopoly, they organized a variety of citizen movements that aimed to eliminate the *latifundios*.

By 1958, 73.3 per cent of the land was in the hands of just 9.4 per cent of the land-holders (Nova, 2002). Further, the *latifundios* held over 4 million ha of land uncultivated, at a time when 200,000 Cuban families were landless and 600,000 people unemployed (Funes, 2002). Illiteracy and disease were also rife, with rural dwellings rarely having electricity, sanitation or fixed running water. The staple foods of the small farmer were rice, beans, roots and tubers. Only 4 per cent of these farmers consumed meat on a regular basis, 11 per cent consumed milk and 20 per cent eggs (Knippers Black et al, 1976; Nova, 2002). Rural and urban lower-income diets were extremely starchy, with a high fat content.[2] The Cuban agricultural worker's weight was said to be 16 pounds (7kg) under the national average, and 35 per cent of the population suffered some form of nutritional deficiency (Knippers Black et al, 1976).

Feelings of insecurity were exacerbated by the monocultural dependency on sugarcane; every price fluctuation on the world sugar market was felt in the Cuban economy, an economy that depended on this crop for 75 per cent of the total value of its exports. Domestically, farm gate prices were low and yet the difference between these and retail prices ranged between 800 and 3200 per cent. At the same time, many imported products could feasibly be grown within the country (Nova, 2002).

The post-Revolution farming system

Agrarian reform

On 1 January 1959, Fidel Castro marched into Havana and declared the Revolution. The US-supported presidential dictator, Fulgencio Batista, fled the country, followed, over subsequent years, by many large landowners (and also by possibly over half the middle class), as the revolutionary government brought in sweeping agricultural reforms (Knippers Black et al, 1976). From the outset, this new government paid great attention to the agricultural sector, as it was the rural agrarian people who had fought for the Revolution. (Even prior to the Revolution, in 1958, Castro's rebel army had brought in an agrarian law stating that farmers who were working state land of less than 62ha should become the owners of that land, and that farmer tenants on private land of less than 26ha should receive the land free of charge.[3])

On 17 May of the same year, the First Agrarian Reform Law was implemented, its main aim being to order the removal of the *latifundios*. It did this by reducing the maximum area allowed to be kept in private ownership to 402ha, while dividing the rest of the land among 200,000 tenant farmers and agricultural labourers. The only exception to this was land dedicated to export crops or to supplying urban areas; in this case the owned area was permitted up to 1342ha. Batista supporters were specifically targeted: their land was confiscated without compensation, and foreigners were not allowed to own sugar factories. At the same time, every farming household with more than five members was supplied with 27ha to live on, and the right to buy up to 67ha. Funes (2002) notes that around this time low-chemical input approaches were encouraged for small farmer production.

The Second Agrarian Reform Law was introduced four years later, in 1963. This changed the structure further, by specifying that all land holdings of more than 67ha would be nationalized, and that all agriculture was to be centrally organized. Many more farmer landowners and professionals fled the country, taking their agricultural knowledge with them. However, many other large landowners responded by dividing their remaining lands into smaller farms amongst close family members and embedding themselves into the peasant economy. In addition, and according to Puerta and Alvarez (1993), they ran down their holdings, which would adversely affect production for years to come. Meanwhile, many farm labourers migrated to the cities.

The free market also disappeared. Farmers had to produce pre-determined crops and sell them to the state, based on an agreed plan. According to Castro, and in line with Marxist-Leninist principles, this system would increase production. Farmers were also required to form cooperatives, in order to facilitate the introduction of mechanized and large-scale production methods and scientific technologies. In this year, nearly 70 per cent of land and 80 per cent of sugarcane volume came under state control (Seraev, 1988).

Prior to the Revolution, only very limited areas had been subject to chemical fertilization, and mechanization was limited to rice and sugar cultivation. Increasing the level of agricultural technology was now a priority of the revolutionary government. New equipment and inputs were imported, supplemented by fertilizer-producing complexes scattered around the country. One of the largest was at Nuevitas, which had an annual production capacity of 200,000 tons of ammonium nitrate, 160,000 tons of nitric acid, 110,000 tons of ammonia and 35,000 tons of urea. Approximately 50 large dams, and more small ones, were constructed during the early post-revolutionary years (Knippers Black et al, 1976).

Political and productive specialization

Politically, Cuba was abolishing market-led capitalism, nationalizing foreign-owned enterprises, and instituting many Soviet-style agrarian and industrial measures. As soon as the First Agrarian Reform Law had been implemented, the USA cancelled its contracts for the purchase of Cuban sugar. The Soviet Union stepped in with a high offer (Murphy, 1999). Relations with the United States deteriorated to a low point with the attempted invasion of the Bay of Pigs in 1961, and the missile crisis of 1962. Subsequently, in 1964, the USA imposed a total trade embargo against Cuba; exports to Cuba had been forbidden since 1960, but imports of Cuban goods were now also banned. Although the missile crisis experience had shown Russia to be untrustworthy, this embargo pushed Cuba to further align itself politically and economically with the Soviet Union and the Socialist Bloc,[4] and the revolutionary government converted into the Cuban Communist Party (Partido Communista de Cuba). As Cuba was unable to sell produce on the world market, it became dependent on the Socialist Bloc for resources it could not, or did not, produce itself: petrol, gas, certain food-stuffs, fertilizers, pesticides and machinery. These resources were purchased largely with the revenue from its sugar, which it continued to sell to the Socialist Bloc on most favourable terms: prices paid were, on average, 5.4 times higher than the world market price. Such a dependent relationship was formally acknowledged in 1972, when Cuba joined the Council for Mutual Economic Assistance (CMEA), the economic organization of communist states.

This political alignment was also apparent in the agricultural sector, which adopted a Soviet-style, industrialized agricultural strategy. At the start of the Revolution, the state had intended to reduce the area of land devoted to sugar and to diversify, in order to better meet its food security needs (Ríos et al, 2001; Nova, 2002). However, a combination of factors prevented this from happening. For one thing, some experimentation in this direction by the state did not see immediate, favourable results (Deere, 1996). The country's economic boom in the 1960s turned out to be brief and based on short-term strategies, while longer-term investment in land declined (Schusky, 1989). Additionally, the strong demand for Cuba's produce from the Socialist Bloc, the need for export revenue, and the ideal natural, social and infrastructural conditions for its production, also contributed to the continued monoculture of sugar (Nova, 2002). This was specifically promoted by President Castro, who invested heavily in the industry in order to achieve record harvests. By the end of the 1960s, domestic production of food had declined at the expense of industrial expansion; sectors were heavily developed for the industrialized production of cattle and dairy, pigs and poultry, rice for national production, and citrus for export.

The further industrialization of Cuban agriculture

Given the hierarchical structure of pre-Revolution agriculture, Cuba found itself with few agronomists or technical specialists, most of them having fled the country. Traditional farmers still existed, but the prevailing political outlook placed greater trust in science and technology than local knowledge, and so Cuba developed a national agricultural system using imported technical expertise mainly from Eastern European countries. This led to the development of modernized, large-scale systems using techniques that had been developed in temperate regions, systems that were similar to, but more intensive than, the previous US-style plantation agriculture. The emerging Green Revolution technologies of the 1960s and 1970s were readily compatible with this approach, with their monocropping over large extensions, and their intensive use of machinery (tractors, harvesters and aviation), chemical fertilizers and pesticides, as well as of livestock feeds. The state applied this approach to all its farms, which by that time covered over 70 per cent of the country's agricultural area, concentrated around the capital. In 1975, 48 per cent of the cultivated area in Havana Province was under monoculture, 34 per cent under 'accompanied' monoculture, and only 18 per cent was relatively diversified (Zequeria Sanchez, 1980). Interestingly, because of the socialist approach, the credit packages of the Green Revolution were not implemented, and thus small farmers did not incur debts to the same extent as in other countries (Funes, 2002). Nevertheless, in the move to eliminate small farming systems and encourage rural-to-urban migration, much traditional knowledge was lost, and the rural population declined from 56 per cent of the total in 1956 to 28 per cent by 1989 (Pérez Marín and Muñoz Baños, 1992).

These decades also saw the 'chemicalization' and 'tractorization' of agriculture (Pérez and Vázquez, 2002; Ríos, 2002). Pest control became based exclusively on chemicals, with predefined norms for each crop. Cultural and traditional control methods were abandoned. With the appearance of new and resistant pests, the State Plant Protection System (INISAV) was created in the mid-1970s, and along with it a cross-country network of early-warning stations. This encouraged a more controlled use of pesticides, and cut their usage by half. By 1982, Integrated Pest Management (IPM) was implemented as official state policy, and with this commenced a small line of research on biological control agents. The Ministry of Sugar (MINAZ) pioneered alternative approaches as early as 1980 through the establishment of a National Programme for Biological Pest Control, and by 1985 had constructed 50 small-scale centres for the production of bio-pesticides for sugarcane. In 1988, the Ministry of Agriculture (MINAG) approved a similar programme (Pérez and Vázquez, 2002). The use of chemical fertilizers was brought somewhat under control when the Ministry of Agriculture established the Directorate of Soils and Fertilizers, which oversaw various soils research institutes. This helped to slow down the trend of deforestation that had existed since pre-Revolution times. Nevertheless, soil degradation continued overall, owing to the effects of intensive chemical use, heavy machinery and large-scale irrigation (Treto et al, 2002). In the 1980s, some research was being undertaken on recycling sugarcane wastes, green manures and the use of bio-fertilizers, but little was practically implemented, save for the production of compost and worm humus, which were applied to sugarcane systems from 1984 onwards. In terms of mechanization, from the time of the Revolution up to 1990, the number of tractors increased from 9000 to 85,000, while oxen numbers decreased from 500,000 to 163,000 over the same period (Ríos and Aguerreberre, 1998). Meat and dairy production was made more efficient in the 1980s through improving pastures and producing feeds from local crop by-products, although these local feeds had difficulty competing with the cheap imports from the Socialist Bloc (Monzote et al, 2002).

Agricultural research also took on the Green Revolution principles, including the concept of biotechnology. Research was highly disciplinary, with each institute specializing in particular crops and commodities. There was little integration, structured planning or evaluation (Bode et al, 1998). For economic and ideological reasons, the national research sector grew to become one of the largest in the Americas, certainly one of the largest compared to those of less-industrialized countries. By 1985 the sector comprised 2500 scientists with an annual budget of $45 million (Casas, 1985). The Ministry of Agriculture combined 17 research centres and 38 experimental stations under its auspices, with additional centres under the Ministries of Higher Education and of Science and Technology (Funes, 2002). In terms of international collaboration, Cuba maintained networks with international organizations during the post-revolutionary period, and particularly with the Food and Agriculture Organization, on whose Council it had sat as a member since 1979. Relations also existed with the Consultative Group on International Research (CGIAR), in particular through the exchange of genetic resources. Within the Caribbean region, Cuba was also active in commodity research and development of crops such as sugar, citrus, maize, bean and potato (ICPPGR, 2000).

By 1985, irrigation covered 26 per cent of total cropped land, largely for sugarcane (46 per cent) but also for rice (20 per cent), citrus (12 per cent) and pastures (10 per cent) (Alvarez, 1994). Chemical fertilizer use amounted to 192kg/ha, and pest and disease controls to 9500 tons (or US$80 million) per annum. At 1989, 1.3 billion tons of fertilizers were being imported, an increase of over 900 per cent since 1958 (Alvarez and Messina, 1992). For pesticides, Cuba was importing 10,000 tons in 1989 (a 200 per cent increase since 1965), along with 17,000 tons of herbicides (a 3300 per cent increase from 1965). Animal feed imports had also risen by 900 per cent since 1965. Overall, 94 per cent of fertilizers, 96 per cent of herbicides and 97 per cent of livestock concentrates were being imported (Funes, 1997). Meanwhile, the value of imports of agricultural machinery had risen by 250 per cent between 1970 and 1989, peaking in 1984, when 90,000 tractors were being used (Díaz, 1999). In 1985, there was an average of one tractor for every 37ha in Cuba, compared with one every 55ha as the global average (Figueroa, 1999b). Box 4.1 provides a description of the intense range of types and source of agricultural inputs and implements, showing that while there was dependence on the Soviet Union and other CMEA members, Western European countries also played a significant role in supplying inputs.

Box 4.1 *Sources of inputs used in Cuba up to 1989*

Until 1989, the majority of the fertilizers came from the USSR, with some quantities imported from the German Democratic Republic and a few Western countries. Herbicides and other pesticides were purchased mainly from Switzerland, both the Federal Republic of Germany and the German Democratic Republic, the United Kingdom and other Western countries. The USSR was the main supplier of animal feed, while the Netherlands sold substantial quantities in several years and minor quantities were imported from the Federal Republic of Germany, France, Peru, Chile and other countries. Agricultural tools came from Spain, Italy, the USSR and Eastern European countries.

   Almost all of the tractors were imported from the USSR, although, in several years, some were purchased in Italy and the German Democratic Republic. All sugarcane loaders and cane wagons were imported from the USSR. Bulldozers were purchased mainly in France, the USSR, Italy and Japan. The vast majority of diggers and cranes were imported from the USSR, although some were purchased in the United Kingdom and the German Democratic Republic. Tractor wagons came mostly from the USSR and, in some years, from Sweden, the United Kingdom and France. Water pumps were imported mainly from the USSR, while Bulgaria, the German Democratic Republic, France, the United Kingdom and Italy sold some quantities to Cuba in later years. Finally, spare parts were mainly provided by the USSR, but some shipments also came from Romania, Bulgaria, the German Democratic Republic and Italy.

Source: Alvarez and Messina, 1992, pp1–2

Overall, Cuban agriculture was applying higher levels of fertilizers, mechanization and irrigation to its fields than the USA, and certainly more than other Latin American countries (Sinclair and Thompson, 2001). All this signified a high level of state agricultural investment: between 1960 and 1990, 25 per cent of all domestic investments went into the agricultural sector (Díaz González, 1999). Yet although the production of certain crops and livestock had more than doubled since the start of the Revolution (Murphy, 1999), production overall was relatively low. The increase in production of selected crops between 1976 and 1989 is shown in Figure 4.2.

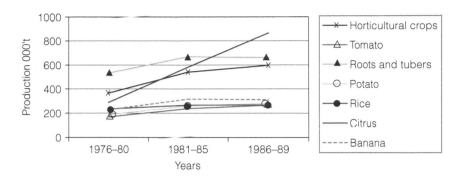

Figure 4.2 *Average annual growth in selected agricultural produce, 1976–1989 (000't)*

Source: Nova González, 1999

Despite this move towards Soviet-style agriculture, in some respects Cuba never became the Caribbean equivalent of the monolithic Eastern European communist system. For example, because Cuba did not force collectivization to the same degree, so some private farmers still remained. However, similarities to the Soviet model did exist in terms of collectivization, centralization and egalitarianism, and to both the Soviet and US models in terms of the large-scale export orientation, subsidies, monocrop focus, high levels of mechanization and dependency on chemical inputs (Gey, 1988; Eckstein, 1994; Mesa-Lago, 1998; Sinclair and Thompson, 2001). Díaz González (1999) notes the relatively low level of productivity and efficiency, and perhaps more importantly, that 80 per cent of agricultural employees – state farm workers – were on salaries lower than those of employees in other economic sectors and that they enjoyed few other incentives. Nevertheless, unlike other non-European communist countries such as North Korea and Taiwan, which systematically invested the proceeds of agriculture into industrialization, Cuba re-invested much agricultural revenue into developing rural infrastructure such as irrigation, transportation, fertilizers, rural education and health (Schusky, 1989). Yet this strategy accentuated Cuba's dependence on agriculture. Cuban officials were aware of this vulnerability from early on, and believing the future to lie in technological expertise they also invested billions of dollars in developing biotechnology, IT, health and robotics (Rosset and Benjamin, 1994).

Types of farming system in the 1980s

Cuban farms were divided into state and non-state ownership. After the agrarian reforms of the Revolution, approximately 30 per cent of agricultural land was owned by individual farmers, and after that time the state made various attempts to integrate these into the centralized and specialized food planning, production and distribution system, by purchasing or renting this land (Zimbalist and Eckstein, 1987; Ramírez Cruz, 1994). By the late 1980s, these efforts had led to four main types of farm organization: state farms, Agricultural Production Cooperatives (CPA), Cooperatives for Credit and Services (CCS), and individual farmers. In 1989, private or non-state production contributed 35 per cent of national production and 48 per cent of export foodstuffs, despite the fact that it covered only 20 per cent of total agricultural area and received lower investments (Nova, 1994; Nova, 2002). Each of these four main types of farm organization is described below.

State farms (fincas estatales)

Covering 80 per cent of cultivable land, state farms were the main producers of staple foods such as rice, milk, beef, poultry, roots and tubers, and of sugar, citrus and coffee. Their activities also covered forestry. These farms received priority over private farms for the receipt of production inputs, technical assistance, credit, investments and new technology, and housing and social services were provided as incentives to workers (Puerta and Alvarez, 1993). They came in various organizational forms and under different ministries. Under the control of the Ministry of Agriculture were non-sugarcane farms and people's farms (*granjas de pueblo*). The Ministry of Sugar controlled sugarcane plantations and agro-industrial complexes (CAIs) (Cole, 1998), while the Ministry of Home Defence also operated production units. All produce went to the state food procurement agency, Acopio.

Non-state farms

Non-state farmers held the major responsibility for subsistence crops such as beans, maize and vegetables, as well as for tobacco and cocoa (Martín, 2002). These farms were characterized by the fact that the land was owned by the farmer or the cooperative rather than by the state. The farmers themselves were classified as peasant farmers or *campesinos*. There were three types of non-state farm, as follows:

Credit and Service Cooperative, CCS (Co-operativa de Credito y Servicio)

Instigated in 1962, the state encouraged independent farmers to organize themselves into cooperatives of between 35 and 75 members, each with an elected management consisting of president, vice-president and administrator. Land was held privately, but assets belonged to the state. Credit relations operated between the bank and the cooperative rather than directly with each farmer, and services offered to the members included the sharing of equipment such as trucks for marketing purposes, or a tractor for cultivation. Produce could be marketed by a cooperative representative. Each CCS member made an individual production plan with Acopio, for some of his or her produce. Not all cooperatives managed to arrange group access to credit and services (Miedema and Trinks, 1998).

Agricultural Production Cooperative, CPA (*Co-operativa de Producción Agropecuaria*)

The second oldest type of production cooperative in Cuba, CPAs were initiated in 1977. In these, farmers were encouraged to pool their land, labour, livestock and materials such as tractors and other tools into collective property, and all investments and production outputs were shared. If land holdings were dispersed, the state could provide the cooperative with the interlying land in perpetuity. This attracted non-landowners and skilled staff to join the cooperatives. Members who pooled their own land and goods or materials were compensated. The cooperative could found a community, with a primary school, small clinic, recreation area and houses, and electricity. CPAs tended to be larger than CCSs, with membership ranging from 50 to 200 per unit. Staff were also more numerous, consisting of a president and heads of production, economics, machinery, veterinary resources, plant health and policy. Members were commonly grouped into work brigades of between 10 and 20 people, each under a brigade head. The staff met on a fortnightly basis, and members every month. Acopio visited regularly to make and oversee production plans with the cooperative staff and heads of brigades, and plans were made for the majority of production. Politics was embodied in the cooperative, with some members being active in the local branch of the National Association for Small Producers (ANAP) or the Cuban Communist Party (PCC). Both membership and productivity of the CPAs declined steadily during the late 1980s, owing to the level of state control that made them unattractive to potential new *campesino* members (Deere et al, 1992; Deere, 1996; Martín, 2002). Agricultural labourers involved in the CPAs held less stake in, and therefore care for, farm performance (Miedema and Trinks, 1998).[5]

Independent farmers (*campesinos independientes, parceleros*)

Despite the efforts of the state to collectivize, these farmers chose to remain independent on their own land and to rely on their own productive capacity. Some received land in perpetuity from the state. They did not fall under any state organizational enterprise, and although they could choose to follow a production plan and sell this produce to the state, this probably made up only a small part of their total production. Access to inputs was controlled, although they made their own investment plans. These farms were small (up to 26ha), mixed, and largely produced for subsistence purposes. The number of independent farmers also declined towards the end of the 1980s (Martín, 2002).

In terms of socio-economic differences between these farm types, comparative studies during this period are rare. One undertaken in 1991 by Deere et al (1995) found that since 1959 farm income levels had improved dramatically, especially for the private sector, while regional differences had lessened over that time. At the same time, all types of agricultural household – whether working individually, in a cooperative or for the state – relied on multiple income sources.[6]

Food collection, distribution and markets

The ration system and planned production
Since 1959, state politics had favoured centralization of the market in order to guarantee equal distribution and stable prices (Benjamin et al, 1984). In 1961, the state food procurement and distribution entity, Acopio, was formed to deal with the majority of food collection and marketing. National wages were raised substantially so that workers could afford more and better foodstuffs. However, this strengthened purchasing power coincided with a decrease in expenses such as rent and electricity, and, coupled with a drop in production during the changes in farm ownership and organization, led to existing food stocks becoming depleted (Alvarez, 2004a). Therefore, in 1962, Cuba established a food ration system to control the sale and flow of food, administered by the Ministry of Internal Trade (MINCIN). This system allowed the population to purchase a set, moderate quantity of basic products (*canasta basica*), at a negligible price. The ration provided most items, such as rice, beans and cooking oil (Rosset and Moore, 1997; Oliveros Blet et al, 1998). As the domestic production of some foodstuffs increased, so these were taken out of the ration system and sold freely, and when they became scarce they were added in again (Alvarez, 2004a). In 1968, as food shortages increased, direct farm sales were partly curtailed, and the specialized production plans instituted whereby the *campesino* farmers would agree with the state to supply a set number and quantity of crops for the ration. This plan included agreement on planting areas and cultivation practices, and planning was undertaken through a designated local state farm, termed 'Agricultural Enterprise', which would supply the farmer with credit, inputs and technical assistance required, such as these Enterprises were already using for their own production activities (Deere, 1996).

By 1971, rationed foods provided a daily quota of 1427 calories. The ration was rather lacking in variety, but did guarantee a basic food security for all. Subsidized food could also be obtained through the canteens of factories and schools. Unrationed food at that time could be obtained through home garden production, restaurants and the black market (Knippers Black et al, 1976). As the major food shortages disappeared toward the end of the 1970s, this situation eased off, and in the 1980s the ration system almost lost its role because of the overabundance of food.

In 1976, the Empresa de Frutas Selectas (Select Fruit Enterprise) was formed, a tourist food delivery organization that provided a direct link between fruit producers and tourist centres. It was able to purchase directly from the farmer rather than through Acopio. In 1979, a Social Security System was established, which included assistance components for the most vulnerable and specifically old people, disabled, single mothers, children and youth. By the 1980s, more than 3 million food portions were being served daily, free or at a very low price, in public institutions such as schools and hospitals (MSP, 1988).

The parallel and campesino markets

One of Acopio's activities was to process imported raw foodstuffs into products for distribution to work places, school canteens, hospitals and old people's homes. This gave rise, in 1982, to what was termed the 'parallel market', a domestic supply of nationally processed products to rival similar, imported ones. This market supplied preserved meat products, dairy and conserved fruit and vegetables, and contributed to increasing food supply as well as stimulating national industrial production.

An additional source of food was the informal *campesino* distribution market, which was useful to supply areas that the state could not reach. Prices on the *campesino* market were higher, but they provided greater diversity and were a major source of meat and dairy products. In the early 1980s, the state experimented by allowing the formation of organized, official *campesino* farmers' markets. According to Oliveros Blet et al (1998), this experiment was stimulated by rising prices in the informal market sector, and by the realization that the state farming sector was inefficient. It was also a way for the state to exert some control over production that was surplus to the agreed plan for Acopio (Figueroa and García, 1984). However, this experiment was unsuccessful. No legislation had been made regarding marketing parameters, and so intermediaries emerged who were buying in bulk and speculating on prices, with no limits on their profit levels. Further, private farmers diverted their resources from planned to market production, leading to a decline in produce destined for the ration system (Deere, 1997). These markets did not therefore lead to the desired decrease in food prices, and a sector of the population was growing disproportionately wealthy. In this sense, these markets were also contradictory to the state goal of the time of encouraging *campesino* farmers to pool their land and resources (Deere and Meurs, 1992). By 1986, these farmers' markets had been shut down by the state. The range of fresh produce then fell, though this was partially substituted by the increasing growth of the parallel market. All commercialization returned to state control, and with this the black market grew.[7]

Dependency on food imports at the end of the 1980s

A major limitation to the food distribution system throughout the 1980s was that, although markets were partially liberalized, the transport system was not, and so products frequently reached the consumer in poor shape. Acopio was notorious for failing to collect produce from farmers, and so wastage was high. Alvarez (2004b) calculates that 10–15 per cent of available food has been wasted in the years since 1980.[8] Further, any attempt at production diversification was hampered by the continued dependency on sugar as the source of foreign exchange (Enríquez, 1994). By the late 1980s, Cuba was still receiving the balance of its food requirements from the Socialist Bloc, including food to be channelled through the ration and social systems and parallel market system, as shown in Table 4.1.

Table 4.1 *Main agricultural imports and their contribution to total national food requirements in Cuba, by the late 1980s (%)*

Crop	%
Wheat	100
Beans	99
Livestock concentrates	97
Oil and lard	94
Cereals	79
Rice	50
Fish	44
Milk and derivatives	38
Poultry	33

Source: Rosset and Benjamin, 1994; Funes, 1997

In 1985, for example, the Ministry of Agriculture was producing just 28 per cent of nationally consumed calories, the rest coming from lands of the Ministry of Sugar and over half from abroad (Casanova, 1994; Garfield, 1999). It is estimated that 57–80 per cent of proteins and 50–57 per cent of calories were supplied by imports (Felipe, 1995; Murphy, 1999). Within the country, 53 per cent of arable land was devoted to export crops, and 44 per cent to national foodstuffs (Pérez Marín and Muñoz Baños, 1992; Enríquez, 2000; Nova, 2002). By 1989, 87 per cent of Cuba's foreign trade was with CMEA (Mesa-Lago, 1998), which was supplying all of Cuba's fuel and 80 per cent of its machinery needs, as well as two-thirds of its foodstuffs. During the final quarter of the 1980s, Cuba started to experience an economic recession (Figueroa, 1999a). Estimates of the amount of Soviet economic aid to Cuba at this time vary between $600 million and $5 billion per year, through price subsidies and low-interest loans (USIA, 1998).

Overall, the food system in operation in Cuba was supplying the needs of the population to a large degree, but the way that it went about achieving this left it vulnerable to external shock, and its greatest weakness was the reliance on a single cash crop. Sugar and its derivatives were providing 75 per cent of Cuba's foreign exchange earnings (Enríquez, 2000), fetching far above world market prices, which themselves were becoming increasingly volatile. This vulnerability was not only economic but also environmental and social. Cuba's crops were adversely affected by climate extremes; in drought-free years, for example, the island was often hit by hurricanes. From the social perspective, labour demand was highly concentrated around the sugar harvest season, and highly technology dependent throughout the year (Schusky, 1989).

Food consumption and health

The Revolution had brought Cubans a National Health System, which by the end of the 1980s included programmes for food-related disease control, and for improving the status of pregnant women and undernourished children. Within the Health System was seated the National Food and Nutrition Surveillance System (SISVAN), which was set up in 1977 with the support of the United Nation Children's Fund (UNICEF) and the Pan American Health Organization (PAHO). The objective of SISVAN was to monitor the population's nutritional status, and it achieved this through surveillances of mother and child nutrition (through the primary health care network), of social feeding, and of chemical and biological food contaminants. It identified and dealt not only with the undernourished but also the obese (Amador and Peña, 1991). The SISVAN system was so successful that it was used as a model by other countries.

Food intake had risen overall during the revolutionary period: in 1989, daily food consumption per capita was relatively high at 2834 calories (including 76g protein), compared to 2500 calories in 1965 (Amador and Peña, 1991; Deere, 1992; Ríos Labrada et al, 2000; Alvarez, 2004b). In the 1980s, Cuba was one of the very few so-called developing countries with life expectancy and mortality rates comparable to developed countries (Garfield, 1999). Similarly, 'Western' diseases were also more prevalent. Mortality rates due to chronic non-communicable disease increased between 1968 and 1989 (Amador and Peña, 1991). This included a 25 per cent increase in cardio-vascular disease, and an almost doubling in diabetes. Obesity in children and young adults also increased. However, apart from a mild iron deficiency, no other chronic nutrient deficiencies were identified in Cuba prior to 1990 (Amador and Peña, 1991).

Food habits were poor and did not improve with improved economic status. In 1989, a study on changing food habits (Muñoz, 1989) identified that almost a quarter of the population could be characterized by their high intake of energy-related foods: animal fats, sugar and rice. As incomes increased, so did the consumption of these foodstuffs. Amador and Peña (1991) also noted that the increased capacity to acquire food had not corresponded with a change in food habits. For example, at that time Cuba had the highest per capita sugar consumption in the world, representing almost 20 per cent of total energy intake. Fresh fruits and vegetables were consumed in low levels and seasonally, and only a low proportion of animal intake was fish. Almost the only vegetables consumed were onions, cucumbers, peppers and tomatoes (García Roché and Ilnitsky, 1986). Amador and Peña (1991) suggest that these dietary habits were the most important contributory factor to increased mortality rates.

Signs of change at the end of the 1980s

That Cuba had never been self-sufficient in the post-revolutionary period was not uncommon – very few countries are – but in terms of national security, Castro (1996) admitted: 'During the years of full economic stability and development of agricultural production, the country reached considerably high levels but not enough to satisfy the needs.' Efforts in the mid-1980s to increase productivity by opening *campesino* farmers' markets had not had the desired outcomes. For this reason, in 1988–89 and prior to the crisis, the state instigated a National Food and Nutrition Programme (*Plan Alimentario*), in order to improve the nutritional status of the population and to increase self-reliance and production. This programme intended to intervene in areas of production planning, imports, marketing, food preparation and distribution. To encourage better food habits, it aimed to decrease sugar consumption to less than 15 per cent of total energy intake, reduce total fat intake and increase the proportion of vegetable fat, fish and fresh fruit and vegetables (ANPP, 1991). It was an ambitious programme. Amador and Peña (1991) illustrate the scope of the programme as involving 'co-ordinating the activities of agriculture and animal production, the food industry, domestic and foreign trade, education and mass-diffusion media, public health, and mass organizations in the framework of the National Food and Nutrition Programme...'

In the mid-1980s, there were signs of awareness in Cuba of the negative influence of agricultural practices on food quality and human health. Although evidence was not widely available, information from the Food Nutrition and Hygiene Institute (INHA) demonstrated a positive relationship between zones of intensive use of nitrogen fertilizers, the incidence of stomach cancer and the increased toxicity of ingested nitrites, as well as a general reduction in the nutritive value of food products (García Roché and Ilnitsky, 1986; García Roché and Grillo Rodríguez, 1991). Towards the late 1980s, nitrate residue levels of domestically produced vegetables were comparable to those reported in East European countries (García Roché and Grillo Rodríguez, 1991). The total daily intake of nitrates was slightly above WHO recommendations, particularly for individuals consuming vegetables, root crops (including potatoes) and rice (García Roché, 1987). Of special concern were the nitrate contents of lettuce and bananas, because these two crops were widely consumed, usually raw, and often given to children. However, residue levels varied enormously between samples, owing to environmental and agricultural variables. García Roché and Ilnitsky (1986) recommended rationalizing chemical fertilizer use at this time. In setting limits to vegetable residues, Cuba looked to standards used by CMEA.

During the 1970s and 1980s, other negative impacts of the industrialized farming model were being seen, including large-scale deforestation, salinization, erosion, compaction and loss of soil fertility (Oro, 1992; Ríos, 2002). At least 70 per cent of Cuba's soils were affected to some degree by erosion, including on the most fertile and relatively flat regions; soil compaction affected 25 per cent of agricultural land; and between 10 and 32 per cent was affected by salinization (Espino, 1992; Sáez, 1997; Díaz-Briquets and Pérez López, 2000). Dam construction led to the contamination of aquifers, and by 1989 agriculture was generating 9Mt of solid residues and 27Mm^3 of liquid residues per year, further contaminating groundwater (Atienza Ambou et al, 1992). Yields of the major commodity crops were also decreasing (Nova, 2002). Castro appears to have been aware of the harmful affects of industrialized agriculture. In a publication on environmental management, he is quoted as saying, 'The ultimate responsibility for the cumulative environmental deterioration in the Third World as a whole belongs to the developed capitalist world... The principle producers of pesticides, fertilizers and other noxious chemical products' (Castro, 1993, p22), and, further, 'It is recognised today that, as a result of the so-called Green Revolution, agriculture became highly dependent upon chemical products, with serious implications for the environment. In addition, this created the conditions for the deterioration of genetic diversity, as a result of cultivating high-yield hybrids' (1993, p31). There was also the realization that the monodisciplinary approach to agriculture was not conducive to self-reliance, and that dependency on inputs should be reduced (Murphy, 1999). In addition, the excessively complex structure of the agricultural research and support sector was being seen to have limited effectiveness (Casas, 1985). This included the centralized food collection and distribution system, which in the 1980s was experiencing further logistical problems (Enríquez, 2000).

Unfortunately, before plans for reform could be further developed, the crisis sprung. As Castro (1993) explained: 'In a critical appraisal of its economic policy carried out toward the end of the 1980s, the government decided, in open consultation, to begin a new development stage with a process of transformations that came to be known as the "Rectification of Errors and Negative Tendencies". Before it had time to complete the new policies, the disappearance of Eastern Europe suddenly took place imposing grave consequences on Cuba's external economic relations.'

The critical years: 1989 to 1994

The crisis

In 1989 and 1990, the transformations in Eastern Europe and the former Soviet Union, and the resulting dissolution of CMEA, had a major impact on the Cuban economy. Figures vary considerably as to the resultant loss. According to Cuban sources (PNAN, 1994), the country lost 47 per cent of its supply of diesel, 75 per cent of its petrol, 45 per cent of its electricity and 78 per cent of its chemical fertilizers, and pest and disease control products. Rosset and Benjamin (1993) generally concur with this, as shown in Table 4.2, although they indicated lower petroleum and slightly lower pesticide losses. US Government sources, however, propose that net petroleum imports fell by only 23 per cent and consumption by 20 per cent between 1989 and 1992, and that the knock-on effect reduced electricity generation by 30 per cent (EIA, 2008).

Table 4.2 *Losses of major agricultural imports after the dissolution of the Socialist Bloc*

Input	1989 imports	1992 imports	Reduction %
Petroleum (Mt)	13.0	6.1	53
Fertilizers (Mt)	1.3	0.3	77
Pesticides (US$)	80.0	30.0	63
Animal feeds (Mt)	1.6	0.5	72

Source: Rosset and Benjamin, 1993

Food imports, which Cuba was dependent on, dropped by over 50 per cent, as did the import of seeds (Enríquez, 1994). The total spent on imports fell by 80 per cent from $8100 million in 1989 to $1700 million by 1993, of which $750 million went on the purchase of fuel and $440 million on basic food needs (Ríos Labrada, 1999).

Because of the reserves and surpluses held in the system – of food, fuel stocks, incoming foreign revenue, medicines and so on, the real effects of these depleted resources only came to a head in 1993–94. They were exacerbated by the extension of the US trade embargo, through the 1992 Torricelli Act (and later the Helms-Burton Act in 1996).[9] During this period, GNP fell by almost 50 per cent from $19.3 to $10.0 billion (Casanova, 1994; Funes, 2002). According to Garfield (1999), sanctions added a 'virtual tax' of 30 per cent on all imported goods, because of the more expensive and distant markets that had to be relied upon. However, information from the Bureau of Inter-American Affairs of the US Department of State puts this figure at only 2–3 per cent

(USIA, 1998), and the CIA attribute the decline in GDP to 'the result of lost Soviet aid and domestic inefficiencies' (ODCI, 1998).

The post-crisis impact

Economic and agricultural decline

The economy bottomed out by 1994, at which point Cuba was operating at one-third of its industrial capacity. At 1992, 10–18 per cent of the labour force was unemployed as a result of the closure of industrial plants and of the return to the country of Cubans who had been posted abroad. Enrolment in schools and universities also decreased due to the poor employment opportunities. The housing deficit increased up to 1.1 million units by 1993. With the increase in price of consumer goods, the average disposable income decreased by 46 per cent, and the gap between the highest- and lowest-paid workers rose at the expense of state-sector workers (Mesa-Lago, 1998). During this period, several reports highlighted the hard economic times facing Cuba, and the accompanying unrest among the population (e.g. Klepak, 1991; Schmid, 1991; Kaufman 1992; Robinson, 1992).

State subsidy levels to agriculture fell by at least half, with estimates ranging between $6.5 million and $13.5 million per year (Hatchwell and Calder, 1995; Nova, 1995). The national manufacture of fertilizers dropped by 72 per cent between 1989 and 1995 (Mesa-Lago, 1998). Stocks were depleted, and mono-cultural cropping systems without the inputs required to sustain them gave very low yields, whilst no petrol was available to drive the machinery. One of the few impact studies at grassroots level at that time, undertaken in three municipalities of Cuba, estimated the supplies of fertilizer to the *campesino* sector to have reduced by 40–50 per cent (Deere et al, 1994). The state's capacity to produce seed fell by 50 per cent (Ríos Labrada, 1999). Manual labour was scarce, partly because of the previous state drive for urbanization, but also because there was little incentive to work in agriculture. The agricultural sector contracted by 10.3 per cent in 1992, 22.7 per cent in 1993 and 4.9 per cent in 1994 (EIU, 2000). Overall, agricultural output dropped by more than half between 1989 and 1994 (Mesa-Lago, 1998, based on FAO index; EIU, 2000).

Agricultural exports fell to record lows (Figueroa, 1999b). Sales of sugar halved from 7000 million tons in 1989 to 3663 million tons in 1993, with a similar halving in price during the same period, as shown in Table 4.3. As a percentage of hard currency earnings, sugar exports dropped from 65 per cent in 1992 to 40 per cent by 1994 (Hatchwell and Calder, 1995). Citrus production also dropped, by 60 per cent (Ríos Labrada, 1999).

At this time domestic production also reached its lowest point. For example, national production of roots, tubers and vegetables peaked in 1992 before falling to its lowest levels by 1994 (Figueroa, 1999b). The drop in production of major non-sugar crops is shown in Figure 4.3.

Table 4.3 *Sugar exports from Cuba, 1989–1993*

Year	Sales (Mt)	Income (M$)	Average price per ton ($)
1989	7119	3920	551
1990	7169	4314	602
1991	6732	2260	336
1992	6081	1220	201
1993	3662	753	205

Source: CEPAL, 1997

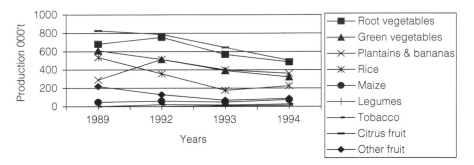

Figure 4.3 *Decline in annual production for major crops between 1989 and 1994*

Source: Figures 1989–96 Economic Commission for Latin America and the Caribbean, based on official Cuban statistics to 1995, in Miedema and Trinks, 1998

Dairy and livestock production was also hard hit because of the loss of imported feed (Frische, 1994). For example, maize imports, largely directed for feed concentrate, fell from 720,000t in 1989 to 95,000t in 1991. In 1992, the country was producing 69 per cent less pork, 89 per cent less powdered milk and 82 per cent fewer chickens than in 1989 (Hatchwell and Calder, 1995). National fish catches halved, decreasing from 1071Mt in 1986–90 to 571Mt in 1991–95, and imports also decreased. These catches went for national consumption, apart from lobster, shrimp and tuna, which were sold to foreign markets (including the domestic tourist market).

Impacts on human health

Although Cuba was still receiving large quantities of food aid (donated or subsidized) from East European and former Soviet countries, in the 1993/94 period average caloric intake dropped by as much as 30 per cent compared to the 1980s (Miedema and Trinks, 1998; Rosset, 2002). Estimates of the magnitude of the decrease vary. Oliveros Blet et al (1998) cite a drop from 2885kcal in

1989 to 2310kcal in 1994, while Figueroa (1999a) estimates an average of 2000kcal/day between 1992 and 1994. Mesa-Lago (1998) was less optimistic, citing a drop from 2845kcal/day in 1989 to 1670kcal/day between 1989 and 1994. This figure is similar to that from the National Office for Statistics, which also provides lower figures for 1994, as shown in Table 4.4. The local FAO office estimated a per capita drop of 40 per cent in protein, 64 per cent in fat and 67 per cent of vitamins A and C between 1989 and 1993 (Díaz González, 1999).

Table 4.4 *Changes in the food consumption at level of macronutrients*

Macronutrient	Year			Recommended level
	1989	1993	1994	
Energy (kcal)	2845	1863	1948	2400
Proteins (g)	77	46	48	72
Fats (g)	72	26	29	75
Carbohydrates (g)	—	362	370	360

Source: ONE, 1997

During the worst years of the crisis, the main, and only, foods available were sugar, rice, roots and tubers. The declining health of street dogs were a visible indicator of food availability, as they traditionally survived on food waste that during this time was all diverted for human consumption. At this point, everyone lost weight and was malnourished.[10] Mortality rose amongst older adults from 48 to 53 between 1989 and 1993 (Mesa-Lago, 1998); mainly due to the lack of medicines to control chronic infections and parasitic diseases – influenza, tuberculosis, diarrhoea,[11] asthma and heart disease. Other age groups were more resilient. The incidence of tuberculosis rose dramatically owing to the poor nutrition, inadequate housing and unsanitary conditions.

Between 1993 and 1994, vitamin deficiency was the cause of a major outbreak of an eye disorder called 'optic neuritis', which affected 45,000 people (Hatchwell and Calder, 1995). About 20,000 people went blind. This epidemic was attributed to the combination of an inadequate diet (specifically essential amino acids) and toxicity – possibly of cyanide from smoking – and was exacerbated by weakened human states through the increase in physical activity.[12] Kirkpatrick (1996) correlates the emergence of this epidemic with the cessation, five months previously, of supplies of food and medicine from the USA. Symptoms included stressed behaviour, skin rashes and obvious weight loss, and although these symptoms were found among both rural and urban populations, *campesinos* with their own land were less affected than state farm workers.[13] The epidemic became a national crisis and the state responded by distributing multivitamin complexes, which appeared to help bring the epidemic under control. One positive outcome was the closer working relationship of the Ministries of Agriculture and of Health.

Inevitably, the Special Period had an impact on food habits. One study in a traditional district of Havana City compared eating habits prior to 1990 with the period 1993–94. It found that the number of people having breakfast and lunch decreased from 90 per cent to 60 per cent, and 83 per cent to 58 per cent, respectively. Almost all the population still had dinner, but snacking decreased from 64 per cent of the sample group to 42 per cent. Other changes included a decline in the incidence of inviting guests for meals, and in family meal reunions (Nunez Gonzales and Buscaron Ochoa, 1995).

The 'parallel market' gave way, as did the networks of work cafeterias and agricultural markets. The majority of the 11,416 state establishments selling prepared food virtually disappeared between 1990 and 1993 (*Granma*, 1995). In the informal and black-market sectors, food prices escalated owing to the devaluation of the Cuban peso. The marketing of food products was left in the hands of Acopio and Frutas Selectas, but due to transport difficulties the food would rot before it reached the consumer (Murphy, 1999). More seriously, much of the food sent from farms never reached the market but was siphoned onto the black market (one-third of food was lost this way in 1994) (Hatchwell and Calder, 1995). Incidence of theft from fields and state warehouses increased. The evolving National Food Programme became focused on achieving a minimum level of self-sufficiency, particularly of roots, tubers and vegetables. It was from this context of disaster and the need to find workable, innovative alternatives that the interest in, and appreciation of, alternative production approaches began to grow.

Summary

The review of events leading up to the collapse of the Socialist Bloc provides a more nuanced understanding of Cuba's historical food and agriculture strategies, as well as the extent of impact of the crisis. On the one hand, Cuba had placed a high priority on ensuring food supply for its population, but the means by which it sourced the food was not sustainable. A large percentage was imported and was dependent on a fragile foreign exchange mechanism. The remainder was produced through a production system that was inappropriate to both the agroclimate and the internal economy, and the consequences for environmental and human health were predictable. The practice of investing the profits of the sugar industry into rural strengthening rather than industrial development may have proved to be both its downfall and its saving grace: there was nothing to fall back on when the sugar income failed, but it was this strong rural capacity that enabled Cuba to survive during the 1990s.

Notes

1   Gross Domestic Product.

2   This poor-quality diet was also out of choice: green vegetables were in low demand, and the abundant tropical fruits were eaten only sparingly (Knippers Black et al, 1976).

3   Puerta and Alvarez (1993) suggest that in those early days only one-tenth of small farmers were actually without some legal claim to land, but it was this small proportion, in the east of the country, which influenced the revolutionary government's early agricultural policies of land redistribution. Subsequent policies and agricultural reforms changed to those of greater state control.

4   Throughout the 1960s and 1970s, Cuba functioned as a Soviet satellite for disseminating its political beliefs to less industrialized countries, encouraging revolutionary movements in Latin America, and sending troops to support causes in Angola and Ethiopia.

5   According to one CPA member, older farmers who could no longer manage their own land had been encouraged to pool it into CPA cooperatives rather than passing it on to their children, and in this way the traditional intergenerational heritage of land, skills and knowledge had been broken.

6   *Campesino* farmers had, according to Lehmann (1985), received an extraordinarily good deal from the Revolution, to the point where the state lost out. These *campesino* farmers gained far larger land units than they might have expected after the Revolution, and opportunity was provided for nuclear families to farm their own plots of land. At the same time they were in receipt of the same food rations as non-food producers, and there was no great pressure on them to sell their produce to the state. That production stagnated is generally felt to be due to the lack of farmer incentives, which exacerbated the country's dependence on imports. However, Burnhill (1985) suggests that the *campesino* sector could in fact have made a larger contribution to agriculture had it not been for state policy on restricting access to resources, and she provides examples where, in practice, the state farm sector outperformed the private sector for this reason. Forster (1982), on the other hand, concludes the opposite: that the private sector has outperformed the state for most crops, although certain crops such as rice and eggs did benefit from large-scale, capital intensive state production. Peña Castellanos and Alvarez (1996) support this view, suggesting that large-scale state farms were unable to provide the detailed attention required for agricultural activities.

7   Figueroa and García (1984) argue that an alternative solution to deal with this non-state marketing would have been to establish 'consumption cooperatives' as were functioning in other socialist countries.

8   To some authors, such as Alonso (1992), the poor state performance over food distribution more than justified the role of the intermediary.

9   Domínguez (1997) argues that the United States would have taken even more direct action had it not been for pressure from its international partners.

10 During interviews amongst health professionals, it emerged that a very small number of deaths could be directly attributed to starvation.

11 The death rate from diarrhoeal disease increased by 250 per cent between 1989 and 1994 (Moore, 1998).

12   For more details, see Gay et al, 1986; Porrata Maury et al, 1995. A similar epidemic that combined poor health with smoking occurred in Cuba in 1896, when Cubans were fighting for independence against Spain and the rural population was forced to move to urban areas, leading to a drop in food production and corresponding poor health. Cuban researchers had noted similar epidemics in some European countries at the end of the Second World War, such as in the Netherlands, amongst British prisoners of war in Japan, and in the Spanish Civil War.

13   According to one source, this was one reason for the subsequent restructuring of farms, in order to improve nutritional supply to the rural population.

References

Alonso, J. F. (1992) 'The farmers' free market: a rejected approach but a possible solution', *Proceedings from the 2nd Annual Meeting for the Study of the Cuban Economy* (ASCE), 15–17 August, Florida International University, Miami

Alvarez, J. (1994) *Cuba's Infrastructure Profile*. International Working Paper Series IW94–4, July, Gainesville, Food and Resources Economics Department, Institute of Food and Agricultural Sciences, University of Florida

Alvarez, J. (2004a) *Overview of Cuba's Food Rationing System*. EDIS Document FE482, Gainesville, Department of Food and Resource Economics, Florida Cooperative Extension Service, University of Florida

Alvarez, J. (2004b) *The Issue of Food Security in Cuba*. EDIS Document FE483, Gainesville, Department of Food and Resource Economics, Florida Cooperative Extension Service, University of Florida

Alvarez, J. and Messina, W. A. (1992) *Potential Exports of Florida Agricultural Inputs to Cuba: Fertilizers, Pesticides, Animal Feed and Machinery*. International Working Paper Series IW92–33, December, Gainesville, Food and Resource Economics Department, Institute of Food and Agricultural Sciences, University of Florida

Amador, M. and Peña, M. (1991) 'Nutrition and health issues in Cuba: strategies for a developing country', *Food and Nutrition Bulletin*, vol 13, no 4, pp311–317

ANPP (1991) *El Programa Alimentaria*, Havana, Asemblea Nacional de Poder Popular

Atienza Ambou, A., García Alvarez, A. and Echevarría Vallejo, O. (1992) *Repercusiones Medioambientales de las Tendencias de Desarrollo Socioeconómico en Cuba*, Havana, Instituto Nacional de Investigaciones Económicas

Benjamin, M., Collins, J. and Scott, M. (1984) *No Free Lunch: Food and Revolution in Cuba Today*, San Francisco, Institute for Food and Development Policy

Bode, M. A. M., Boza, A. M. and de Souza Silva, J. (1998) *Instituciones Sostenibles para el Desarrollo Sostenible. El Caso de SINCITA de Cuba*. Discussion Document No. 98-8, The Hague, ISNAR

Burnhill, L. (1985) *The Private Sector in Cuban Agriculture 1959–1985: A Socio-Economic Study*. Occasional Paper 8, Central American and Caribbean Program, SAIS, John Hopkins University, December

Casanova, A. (1994) 'La economía de Cuba en 1993 y perspectivas para 1994', *Boletin informativo*, *CIEM* no 16 (July), pp3–14

Casas, J. (1985) *Cuba: A Small Country, a Large Agricultural Research Potential*. Workshop on Agricultural Research Policy, Minneapolis MN (April)

Castro, F. (1993) *Tomorrow is Too Late. Development and the Environmental Crisis in the Third World*, Melbourne, Ocean Press

Castro, F. (1996) Speech at UN World Food Summit, 18 November

CEPAL (1997) *La Economia Cubana. Reformas estructurales y desempeno en los noventa.* Anexo Estadistico, Mexico, CEPAL

Cole, K. (1998) *Cuba: from Revolution to Development*, London, Pinter

Deere, C. D. (1992) *Socialism on One Island? Cuba's National Food Program and its Prospects for Food Security*. Working Paper Series No. 124, The Hague, Institute of Social Studies

Deere, C. D. (1996) *The Evolution of Cuba's Agricultural Sector: Debates, Controversies and Research Issues*. International Working Paper Series IW96-3, Gainesville, Institute of Food and Agricultural Sciences, Food and Resource Economics Department

Deere, C. D. (1997) 'Reforming Cuban agriculture', *Development and Change,* vol 28, pp649–669

Deere, C. D. and Meurs, M. (1992) 'Markets, markets everywhere? Understanding the Cuban anomaly', *World Development*, vol 20, no 6, pp825–839

Deere, C. D., Meurs, M. and Pérez, N. (1992) 'Toward a periodisation of the Cuban collectivization process: changing incentives and peasant response', *Cuban Studies/Estudios Cubanos*, vol 22, pp115–149

Deere, C. D., Pérez, P. and Gonzales, E. (1994) 'The view from below: Cuban agriculture in the "Special Period in Peacetime"', *The Journal of Peasant Studies*, vol 21, no 2 (January), pp194–234

Deere, C. D., González, E., Pérez, N. and Rodríguez, G. (1995) 'Household incomes in Cuban agriculture: a comparison of the state, co-operative and peasant sectors', *Development and Change*, vol 26, no 2 (April), pp209–34

Díaz-Briquets, S. and Pérez-López, J. (2000) *Conquering Nature – The Environmental Legacy of Socialism in* Cuba, Pittsburgh PA, University of Pittsburgh Press

Díaz González, B. (1999) 'Collectivisation of Cuban state farms: a case study', in Lara, J. Bell (ed) *Cuba in the 1990s*, Havana, Editorial José Marti

Domínguez, J. I. (1997) 'US-Cuban relations: from the Cold War to the colder war', *Journal of Inter-American Studies and World Affairs*, vol 39, no 3 (Fall), pp49–75

Duran, J. L. (1998) 'Degradación y manejo ecológico de los suelos tropicales, con énfasis de los de Cuba', *Agricultura Orgánica*, vol 4, no 1 (April), pp7–9

Eckstein, S. E. (1994) *Back From the Future: Cuba Under Castro*, Princeton NJ, Princeton University Press

EIA (2008) *Cuba Energy Profile*, Energy Information Administration. http://tonto.eia.doe.gov/country accessed in March

EIU (2000) *Country Report: Cuba*, London, The Economist Intelligence Unit

Enríquez, L. J. (1994) *The Question of Food Security in Cuban Socialism*, Berkeley, University of California

Enríquez, L. J. (2000) *Cuba's New Agricultural Revolution. The Transformation of Food Crop Production in Contemporary Cuba*, Development Report No. 14, Dept. Sociology, University of California. www.foodfirst.org/pubs/ devreps/dr14.html

Espino, M. D. (1992) 'Environmental deterioration and protection in Socialist Cuba', *Cuba in Transition*, vol 2, pp281–292

Felipe, E. (1995) 'Apuntes sobre el desarrollo social en Cuba', *Boletin Informativo. CIEM,* no 20 (March–April), pp3–16

Figueroa, V. (1999a) 'Revolución agraria y desarrollo rural en Cuba', paper presented at *Conference on Co-operativism*, October, Santa Clara, UCLV

Figueroa, V. M. (1999b) *Cuba: de la Recesión al a Crisis y Reforma del Modelo Económico de la Transición*, Villa Clara, UCLV

Figueroa, V. and García, L. A. (1984) 'Apuntes sobre la commercialización agrícola no estatal', *Economía y Desarrollo*, vol 83, pp34–61

Forster, N. (1982) 'Cuban agricultural productivity: a comparison of state and private farm sectors', *Cuban Studies/Estudios Cubanos*, vol 11, no 2/12/2 (July 1981–January 1982), pp105–125

Frische, K. (1994) 'The crisis in Cuba continues', *Aussenpolitik*, vol 45, no 3/94, pp299–305

Funes, F. (1997) 'Experiencias Cubanas en agroecologia', *Revista Agricultura Organica* (August–December), pp10– 14

Funes, F. (2002) 'The organic farming movement in Cuba', in Funes, F., Garcia, L., Bourque, M., Pérez, N. and Rosset, P. (eds) *Sustainable Agriculture and Resistance: Transforming Food Production in Cuba*, California, Food First Books, pp1–26

García Roché, M. O. (1987) *Ernahrung*, 11(610) (manuscript obtained in Cuba)

García Roché, M. O. and Grillo Rodríguez, M. (1991) 'Limites de residuos permisibles de nitratos en los productos vegetales de Cuba', *Revista CENIC Ciencias Biológicas*, vol 22, pp1–2

García Roché, M. O. and Ilnitsky, A. (1986) 'Contendio de nitrato en productos vegetales cubanos en relación a la ingestión de nitratos por la población', *Revista De Agroquímica y Tecnologia de Alimentos*, vol 26, no 1, pp115–122

Garfield, R. (1999) *The Impact of Economic Sanctions on Health and Wellbeing*. Relief and Rehabilitation Network, Network Paper 31, London, ODI

Gay, J., Grillo, M., Castro, A. and Plasencia, D. (1986) 'Sistema de vigilancia alimentaria y nutricional en Cuba: desarrollo y perspectives', in Gay, J. (ed) *Memorias del Taller Internacional sobre Vigilancia Alimentaria y Nutricional*, Havana, INHA, pp4–21

Gey, P. (1988) 'Cuba: a unique variant of Soviet-type agriculture', in Wadekin, K. E., (ed.) *Communist agriculture. Farming in the Far East and Cuba*, London, Routledge

Granma (1995) 'Resumen del encuentro de presidentes de las asembleas municipales del poder popular', *Periodico Granma* (21 September)

Hatchwell, E. and Calder, S. (1995) *Cuba in Focus: A Guide to the People, Politics and Culture*, London, Latin America Bureau

ICPPGR (2000) 'Cuba CR. Colaboración Internacional', *International Conference and Programme on Plant Genetic Resources*, Rome, IPGRI. http://web.icppgr.fao.org/oldsite/CR/CR/CUBA/c6.htm

Kaufman, S. (1992) 'Collapsing Cuba', *Foreign Affairs*, vol 71, no 1, pp130–135

Kirkpatrick, A. F. (1996) 'The role of the USA in the shortage of food and medicine in Cuba', *Lancet*, no 348, pp1489–1491

Klepak, H. (1991) 'Hard times ahead for Cuba', *Jane's Defense Weekly* (12 October), pp666–668

Knippers Black, J., Blutstein, H. I., Edwards, J. D., Johnston, K. T. and McMorris, D. S. (1976) *Area Handbook for Cuba*, Washington DC, Foreign Area Studies, the American University

Lehmann, D. (1985) 'Smallholding agriculture in revolutionary Cuba: a case of under-exploitation?' *Development and Change*, vol 16, pp251–270

Martín, L. (2002) 'Transforming the Cuban countryside: property, markets and technological change', in Funes, F., García, L., Bourque, M., Pérez, N. and Rosset, P. (eds) *Sustainable Agriculture and Resistance. Transforming Food Production in Cuba*, Oakland CA, Food First Books, pp57–71

Mesa-Lago, C. (1998) 'Assessing economic and social performance in the Cuban transition of the 1990s', *World Development*, vol 26, no 5, pp857–876

Miedema, J. and Trinks, M. (1998) *Cuba and Ecological Agriculture. An Analysis of the Conversion of the Cuban Agriculture towards a more Sustainable Agriculture.* Unpublished MSc thesis, Wageningen Agricultural University

Monzote, M., Muñoz, E. and Funes Monzote, F. (2002) 'The integration of crops and livestock', in Funes, F., García, L., Bourque, M., Pérez, N. and Rosset, P. (eds) *Sustainable Agriculture and Resistance: Transforming Food Production in Cuba*, Oakland CA, Food First Books, pp190–211

Moore, M. (1998) 'Cuba's health care system haemorrhages,' *Guardian Weekly* (5 April), p15

MSP (1988) *Situación nutricional del país.* Informe Técnico, Havana, Ministerio de Salud Pública

Muñoz, E. (1989) *Análisis tipológico de consumidores de productos alimenticios en Cuba*, Havana, ICIODI

Murphy, C. (1999) *Cultivating Havana: Urban Gardens and Food Security in the Cuban Special Period.* Unpublished Masters thesis, Fac. Latinoamericana de Ciencias Sociales, Universidad Nacional de Habana

Nova, A. (1994) *Cuba: Modificación o Transformación Agrícola.* Mimeograph, Havana, INIE

Nova, A. (1995) 'Mercado agropecuario: factores que limitan la oferta', *Revista Cuba-Investigación Económia*, INIE, no 3 (October), Ciudad de la Habana

Nova, A. (2002) 'Cuban agriculture before 1990', in Funes, F., Garcia, L., Bourque, M., Pérez, N. and Rosset, P. (eds) *Sustainable Agriculture and Resistance: Transforming Food Production in Cuba*, Oakland CA, Food First Books, pp27–39

Nova González, A. (1999) *Nuevo Sistema Agroecológico Productivo. Segunda Quincena – No. 20 Octubre 1999*, Santa Clara, UCLV

Nuñez Gonzales, N. and Buscaron Ochoa, O. (1995) 'Algunas apuntos sobre el sistema alimentario en el barrio de Cerro, Ciudad de la Habana', *Revista Cubana de Alimentación y Nutrición*, vol 9, no 1, pp10–15

ODCI (1998) *Cuba. Factbook of the CIA.* www.odci.gov/cia/publications/factbook/cu.html

Oliveros Blet, A., Herrera Sorzano, A. and Montiel Rodríguez, S. (1998). *El Abasto Alimentario en Cuba y Sus Mecanismos de Funcionamiento.* Unpublished manuscript. Faculty of Geography, University of Havana

ONE (1997) *Estadisticas agropecuarias. Indicadores sociales y demográficos de Cuba*, Havana, Oficina Nacional de Estadísticas

ONE (2006) *Annuario Estadístico de Cuba. Edición 2007*, Havana, Oficina Nacional de Estadisticas

Oro, J. R. (1992) *The Poisoning of Paradise: Environmental Pollution in the Republic of Cuba*, Miami, Endowment for Cuban American Studies

Peña Castellanos, L. and Alvarez, J. (1996) 'The transformation of the state extensive growth model in Cuba's sugarcane agriculture', *Agriculture and Human Values*, vol 13, no 1 (Winter)

Pérez, N. and Vázquez, L. L. (2002) 'Ecological pest management', in Funes, F., García, L., Bourque, M., Pérez, N. and Rosset, P. (eds) *Sustainable Agriculture and Resistance. Transforming Food Production in Cuba*, Oakland CA, Food First Books, pp109–143

Pérez Marín, M. and Muñoz Baños, E. (1992) 'Agricultura y alimentación en Cuba', *Agrociencia, serie Socioeconómica*, vol 3, no 2 (May–August)

PNAN (1994) *Statistics*, Pan American Health Organisation. www.paho.org

Porrata Maury, C., Abreu Penate, M., Hernandez Triana, M., Gay Rodriguez, J., Hervia Perdomo, G. and Marquez Rodriguez, H. (1995) 'Asociacion de la ingestion de aminoacidos esenciales y la neuropatia epidemica en la Isla de la Juventud', *Rev.Cubana deAlimentación y Nutrición*, vol 1 (January–June), pp16–22

Puerta, R. A. and Alvarez, J. (1993) *Organisation and Performance of Cuban Agriculture at Different Levels of State Intervention*. International Working Paper Series, IW93–14 (July), Gainesville, Food and Resource Economics Department, Institute of Food and Agricultural Sciences, University of Florida

Ramírez Cruz, J. (1994) 'El sector co-operativo en la agricultura Cubana', *Cuba Socialista*, vol 11 (June–July), pp1–24

Ríos, A. (2002) 'Mechanisation, animal traction and sustainable agriculture', in Funes, F., Garcia, L., Bourque, M., Pérez, N. and Rosset, P. (eds) *Sustainable Agriculture and Resistance: Transforming Food Production in Cuba*, Oakland CA, Food First Books, pp155–163

Ríos, A. and Aguerreberre, S. (1998) 'La tracción animal en Cuba', *Conferencia en el Evento Internacional de Agroingeniería*, Instituto de Investigaciones de Mecanización Agropecuaria (IIMA), Havana

Ríos, H., Soleri, D. and Cleveland, D. (2001) 'Conceptual changes in Cuban plant breeding in response to a national socioeconomic crisis: the example of pumpkins', in CABI, *Farmers, Scientists and Plant Breeding. Integrating Knowledge and Practice*, Wallingford, CAB International

Ríos Labrada, H. (1999) *Participatory Plant Breeding in Cuba*, Havana, INCA

Ríos Labrada, H., Funes, F. and Funes Monzote, F. (2000) *Alternative Food Production in Cuba: Strategies, Results and Challenges*. Unpublished paper, Havana

Robinson, L. (1992) 'Can the revolution survive? Castro's Cuba is isolated, and its people are facing growing hardships', *US News and World Report*, June 1992, pp41–43

Rosset, P. (2002) 'Lessons of Cuban resistance', in Funes, F., García, L., Bourque, M., Pérez, N. and Rosset, P. (eds) *Sustainable Agriculture and Resistance: Transforming Food Production in Cuba*, Oakland CA, Food First Books, ppxiv–xx

Rosset, P. and Benjamin, M. (1993) *Two Steps Forward, One Step Backward: Cuba's Nationwide Experiment with Organic Agriculture*, San Francisco CA, Global Exchange

Rosset, P. and Benjamin, M. (1994) *The Greening of the Revolution. Cuba's Experiment with Organic Farming*, Melbourne, Ocean Press

Rosset, P. and Moore, M. (1997) 'Food security and local production of biopesticides in Cuba', *LEISA Magazine*, vol 13, no 4 (December), pp18–19

Sáez, H. R. (1997) 'Resource degradation, agricultural policies and conservation in Cuba', *Cuban Studies*, no 27, pp40–67

Schmid, R. A. (1991) 'Despair and protest in Cuba', *Swiss Review of World Affairs*, pp5–16

Schusky, E. L. (1989) *Culture and Agriculture, An Ecological Introduction to Traditional and Modern Farming Systems*, Westport CT, Greenwood Publishing Group, pp47–170

Seraev, S. (1998) *La Transformación Socialista de la Agricultura en Cuba*, Editorial Progreso Moscu.

Sinclair, M. and Thompson, M. (2001) *Cuba Going Against the Grain: Agricultural Crisis and Transformation*, Oxfam America. http://www. oxfamamerica.org/newsandpublications/publications/research_reports/art1164.html accessed May 2002

Treto, E., Garcia, M., Martínez Viera, R. and Manuel Febles, J. (2002) 'Advances in organic soil management', in Funes, F., García, L., Bourque, M., Pérez, N. and Rosset, P. (eds) *Sustainable Agriculture and Resistance. Transforming Food Production in Cuba*, Oakland CA, Food First Books, pp164–189

USIA (1998) *US Regrets Cuban Rejection of Emergency Food Relief* (1 October). www.usia.gov/regional/ar/us-cuba

Weeks, J. M. and Ferbel, P. J. (1994) *Ancient Caribbean. Research Guides to Ancient Civilizations*, New York, Garland Publishing

Zequeria Sanchez, M. (1980) *Uso de la Tierra en la Provincia de la Habana*. Tesis de Grado Doctorado Facultad de Geografia, Universidad de la Habana

Zimbalist, A. and Eckstein, S. (1987) 'Patterns of Cuban development: the first twenty-five years', *World Development*, vol 15, no 1, pp5–22

Life After the Crisis: The Rise of Urban Agriculture

Life after the crisis

The demise of the Socialist Bloc led to a drastic drop in the subsidies received by Cuba. The state had to immediately implement various reforms, and society had to cope. Shortages of imported resources affected all aspects of Cuban life, from transport to health. Resource availability dwindled at varying speeds – depending on how much of the resource had been stockpiled, or was made a priority for importing, or could easily be substituted for. For many months, rolling into years, the situation was grim and frustrating. This was not simply because the low resource levels were grinding systems almost to a halt. Cuban society had been accustomed to the affluent years of the 1980s, the population was highly educated and there was a strong work ethic. Suddenly, they became ineffective. Transport was particularly hard hit, as available oil was prioritized instead for electricity generation. The number of buses in Havana dropped from 2500 to just 500, and it could take four hours to get to work by public transport. Paper ran out, so journals ceased publication. Because of the state planning and management system, structures and processes were not in place to support individual efforts to address and resolve challenges. The day-to-day struggles of Cuban society during the early years of the crisis have been little documented; to a large extent these experiences are of a national personal nature and not the business of the outside world, although Cubans are happy to chat about it anecdotally, and no doubt reflective, first-hand accounts will ensue.

With regard to food and agriculture, the state prioritized spending on food imports and on the replacement of imports with domestic production, and therefore some aspects were less affected than others (Castro, 1996). Over these critical early years, the country found resourceful ways to cope. The basis for this re-emergence was the well-developed social infrastructure and human resource base in which the state had invested since the Revolution (Rosset and Benjamin, 1994; Garfield, 1999; Funes, 2002). This enabled the continuation of primary, subsistence occupations, and in particular agriculture. Importantly, this infrastructure worked in a unified manner to fulfil social objectives; there was no private sector within the country to dilute or divert these efforts by operating towards different objectives. In addition, Cuba was not completely isolated – certain countries did continue their trading and support.

The lack of resources soon instigated recycling and energy-efficient campaigns: the use of bicycles and larger public buses (*camelos*), a national school campaign to collect recyclable materials in exchange for school supplies, active neighbourhood recycling centres, new factories to produce domestic items from recycled materials, and a nationwide environmental education campaign. There has been a tendency to interpret this phenomenon as a proactive choice by Cuban authorities to 'go green', and no more so than through the major post-crisis success story: the rise of urban organic agriculture.

The rise of urban agriculture: the vanguard for a localized organic farming system?

The success of urban agriculture in Cuba provides the basis for many foreign accounts of Cuba's transition towards local organic agriculture. Certainly, Cuba's urban agricultural movement encapsulates all the positive forces arising out of the ashes of the crisis. Because this aspect of Cuban agriculture has already been relatively well documented, its main features are outlined here.

The historical development of urban agriculture

Emergence from the grassroots
Urban agriculture emerged spontaneously out of the hardships of the early 1990s. For urban dwellers who had migrated from the countryside, cultivating urban waste land and keeping small livestock was a natural survival strategy. Possibly the first coordinated effort was the Santa Fe project in the north-west of Havana City, initiated in 1991 by individuals who went on to become co-founders of the Cuban organic movement. Taking advantage of the available resources within the community, they reclaimed empty urban space for food production to help overcome irregular and inadequate food supplies, using the principles of organic agriculture by default. Production tripled over the first three years to supply approximately 30 per cent of local food needs (Windisch, 1994).

Successful state backing
The advantages of, and rationale for, supporting urban agriculture quickly became clear. Urban centres had the highest demand for foodstuffs, especially perishables that were difficult to transport. Such produce – vegetables, fruit, flowers, spices and small livestock – required a high labour input that was readily available in the cities. Further, city wastelands and neglected areas were

otherwise becoming breeding grounds for disease (Companioni et al, 2002). The state recognized the potential of urban agriculture in contributing to the new National Food and Nutrition Programme (of 1989) and supported it by making land available for growers, providing them with appropriate extension services and organizing marketing (Murphy, 1999). In 1994, the Department of Urban Agriculture was established which became, in 1998, part of the Ministry of Agriculture. According to Companioni et al (2002), urban production was based on three principles: the use of organic methods that did not contaminate the environment; the use of local resources; and the direct marketing of produce. Already by 1994, urban production was well developed, and this new department, and research groups allocated to work on urban systems, found themselves running to keep pace with urban producers. As one researcher described it, 'Development was ahead of research due to the high demand for techniques, so it was participatory and spontaneous from the start – we had to give out technologies before we could even test them, and in fact the farmers tested them.'

The urban agricultural movement differed in its development from the more prescriptive, top-down model of agricultural support customarily employed in Cuba; in this case the grassroots was driving research and policy. This was for a very pragmatic reason under the challenging circumstances: urban production showed to double or triple each year and even whilst technologies were still being developed and improved. This immediate success gave the state further incentive to continue with its support and facilitation rather than instigating legislative restrictions. It formalized land access and legalized the right to sell produce. In one example, the governing council of Havana, which included representatives from the Department of Urban Agriculture, turned down a proposal for a joint-venture building planned on an agricultural site in the city. In 1997, a Resolution (527/97) was passed that allowed each urban dweller in Cuba to be eligible for up to one-third of an acre of land. By December 1999, more than 190,000 lots had been taken up (Sinclair and Thompson, 2001). Success could be measured not only in terms of production, but also by the satisfaction of city dwellers who were empowered to resolve their own food problems. The hands-off, facilitatory approach of the state encouraged a diverse range of responses to the heterogeneous local conditions (Companioni et al, 2002). As a member of the Department of Urban Agriculture put it, 'Our objective is for producers to fulfil their production goals with as much ease as possible.' By 1999, urban agriculture was contributing 5 per cent of Cuba's total domestic production, mainly salad crops, according to one researcher. Significant support was also permitted by the state from international sources; in the early days from foreign NGOs interested in supporting the Cuban cause; and later on from bilateral donors concerned with reducing urban poverty. This meant that Cuban researchers had comparatively good access to international knowledge and materials. It also enabled alternative agriculture networks such as the permaculture and organic movements to make significant input into urban agriculture as a serious development option. Sociological studies were also undertaken which may not have been possible in rural areas, such as the work of Murphy (1999) on agriculture and food security, and that of Carrasco et al (2003) on the Agricultural Knowledge and Information System. Finally, this international interchange encouraged foreign reporting to focus on the urban situation.

Urban farming systems

Three main types of urban farm emerged: the state or cooperative owned *organoponicos*; privately owned, intensive community or home gardens (*huertos intensivos*); and parcel plots (*parceleros*).

Organoponicos

These consisted of constructed beds filled with soil and stones for drainage, and then with 40–50 per cent organic material. This raised-bed design was necessary due to the poor quality of the underlying urban terrain. Seeds were generally transplanted from nursery beds. In 1997, there were 400 *organoponicos* in Havana City alone. High-performing *organoponicos* were selected for preferential access to inputs and credit, which enabled them to install high-tech facilities such as cement pavements, microjet irrigation systems, water pumps and iron-and-mesh frames for the protection of crops from the summer sun.

Intensive home gardens

These were smaller, privately owned plots, often worked with family labour and having soils of adequate quality for intensive production without the need for raised beds. Seeds could be planted directly into the soil. By the end of 1997, *organoponicos* and intensive home gardens occupied 1000ha, with 15,000 full-time employees nationwide (1ha supporting an average of 15 full-time jobs). To give an idea of the speed at which the state could roll out effects, its plan was to cover 3300ha by the end of 1998.

Parcel plots

Parceleros were independent from the state, farming on wasteland and sites of collapsed buildings. By the end of the decade there were over 100,000 parcel plots, producing more than *organoponicos* and intensive gardens combined. These *parceleros* were organized into 900 self-help gardening clubs, pooling resources and expertise and organizing workshops and events (Chaplowe, 1996).

Other sites

Three other variants of urban agriculture consisted of self-provisioning plots in the workplace, suburban farms and patio cultivation. Many work and educational buildings organized agricultural production on adjacent land, including 300 organizations in Havana (Companioni et al, 2002). Even the high-rise headquarters of the Ministry of Agriculture overshadowed a self-provisioning cabbage patch for staff lunches. Around the city of Havana were 2000 private and 285 state suburban farms, typically 2–15ha in size. These also occurred around other cities and towns. Private apartment or patio cultivation – where there was no garden – relied on household ingenuity through planting in small containers, on balconies and so on, to create a growing space in a concrete environment. Here could be found small livestock production which required little space and brought-in feed.

Labour force and wages

Salaries depended on the type and organizational structure of each individual unit. To encourage responsibility, employees of state and cooperative-run units received dividends on a profit-share basis. Generally, salaries ranged from US$12–40 per month (245 to 800 Cuban pesos),[1] although this varied. The head of one *organoponico* claimed to earn $100/month, while the owner of a family-run home garden in Havana, working seven days a week, claimed to earn more than the *organoponicos.*

Many urban farmers were retired people, and only a minority had much previous experience in the agricultural sector – the majority were labourers, masons, mechanics, housewives and professionals (Weaver, 1997; López, 2000). This lack of experience or training may, paradoxically, have assisted with the take-up of organic approaches. As one researcher pointed out:

> We believe that one reason for the success of urban agriculture has been that the people involved have not been previously trained in agricultural techniques, and so they have been able to take up ecological techniques more rapidly and efficiently. They are not prejudiced by conventional knowledge and so don't require convincing.

In the 1990s, it was estimated that the urban agriculture sector created 160,000 jobs (López, 2000).

Production targets

By the late 1990s, production plans were more formalized and each of the main types of urban farm had to meet planting quotas for certain crops based on population numbers. The aim was for each person to have local access to a target of 170g/day of fresh produce, without this having to pass through the ration system. Above this quota, the producer was free to plant whatever he or she chose. Urban agriculture produced largely salad crops and other vegetables, with a smaller amount of medicinal herbs, ornamental plants and flowers. Lettuce covered the largest area, up to 60 per cent, as demand was high, but the farmer still had to remain diversified to some extent. By 1999, *organoponicos* and intensive home gardens were producing 215g/capita per day of fresh horticultural crops (MINAG, 1999).

Markets

Parceleros could sell their produce in the street or from their plot. *Organoponicos*, although producing a minimum quantity for the state, were also allowed to sell at the farm gate. Prices were set by the manager of the *organoponico*, but had to fall somewhere between those of the state markets at the lower end, and be 30–50 per cent cheaper than prices in the new, official farmers' markets. By the end of 1999, 505 vegetable stands had emerged in Havana. Modest taxes were imposed on these sales. According to an Oxfam survey in 1997, customers tended to be elderly and the majority spent less than five cents (two Cuban pesos) at these stands.

The use of agrochemicals in urban agriculture

Whilst the emphasis on organic was undoubtedly by default at first, several authors have stated or implied that it was forbidden in Cuba to use agrochemicals in urban agriculture because of the rational concern that chemical usage in the urban environment would be harmful to human health (ACAO, 1995; INIFAT, 1995; Weaver, 1997; Miedema and Trinks, 1998; Ritchie, 1998; Sinclair and Thompson, 2001; Companioni et al, 2002). One foreign information leaflet claimed, for example, that 'Havana is the first city in the world to declare its agriculture fully organic – by law!' (HDRA, 1997). Roycroft-Boswell (2002, p25) states that, 'By 1996, by-laws in Havana allowed for only organic methods of food production.' Chaplowe (1996), on the other hand, reports that chemicals were used, albeit in very small quantities. Those working in the sector at the end of the decade provided more nuanced explanations. The head of one organoponico used agrochemicals when necessary, and he received financial assistance in order to purchase the inputs that required dollar payments. Other urban farmers acknowledged the use of low levels of Formulin, Carbaryl, Thiodan and others (Kerry, 2001; Taboulchanas, 2001). A researcher in the National Urban Agriculture Management and Technology Group provided more detail:

In urban agriculture, our National Group, organized by MINAG and headed by INIFAT, makes unwritten recommendations and a list of standards. Whilst there is no fine to pay if these standards are not met, the producer may not receive continued support from the municipality. In the *organoponicos*, it is recommended not to use chemical fertilizers. Some producers still use them, but the best plots do not. Sanidad Vegetal, with support by CITMA, implements a national standard for urban agriculture which prohibits the use of agrochemical pesticides unless authorized by the local urban farm centre [*granja*]. Checks are maintained by neighbours who may report anyone using such inputs, and the offending producer may have to pay a fine to Sanidad Vegetal.

Overall, some types of urban agriculture were under tighter control than others in terms of the use of agrochemicals.

Organic production strategies

Notwithstanding the sporadic use of agrochemicals, the urban production system generally operated along intensive organic lines. Good management and planning was considered crucial. According to urban producers, much emphasis was placed on the planning stage, as experience had proved this to be the critical factor for efficiency and success; the beds should be always full, and harvests consecutive, with seedlings ready to be planted out when a bed became empty.

Urban agriculture was highly suitable for organic input use. The system was already labour intensive so that micro-management and observation were possible, and both the urban environment and the surface of the raised beds (which were relatively dry and hot) were not conducive to common plant pest habitats. *Organoponicos* received weekly check-up visits by specialists from Sanidad Vegetal, and care was taken to prevent the spread of pests and diseases from rural areas. Miedema and Trinks (1998) found a range of eight biological control products being used in *organoponicos* in Havana City. As well as these biological inputs, other control methods employed included the use of repellent crops such as *Tagetes minuta* (*flor de muerto*, Mexican marigold), catch crops which attracted the pests away from the cash crop, the selection of pest-tolerant crops and varieties, use of clean planting materials, use of nematode-free organic matter, soil inversion to expose potential breeding grounds to the sun, the removal of crop residues, and rotation when possible. The planting of neem trees was also encouraged, for use as a natural insecticide.

One guiding principle for urban agriculture was to 'systematically apply organic matter by using all available local alternatives, and to systematically develop local programmes to assure adequate supplies of organic matter' (Companioni et al, 2002). Organic wastes came from four main sources: animal wastes, plant residues, industrial wastes and residential wastes (Altieri et al, 1999). The main organic materials were *biotierra* (composted sugarcane residues), *gallinasa* (chicken manure and rice chaff) and cow manure, all supplied by state farms. Occasionally, Azobacter[2] was applied as a biofertilizer. In 2000, 69,400 tons of compost were applied, and 80,000 in 2001, at an average rate of 13.5t/m^3 (González Novo and Merzthal, 2002; FAO, 2003).

The timely availability of seeds was immensely important. Cuban varieties were used where possible as these were adapted to local conditions, and most urban farms were self-sufficient in reproducing and saving seeds from certain crops. However, many of the salad crops produced did not seed in the hot climate and so new stocks had to be imported each year. The State Seed Enterprise was unable to meet urban agricultural demand, especially at peak periods, so in 1998 a network of provincial seed farms was set up. In the urban environment it was observed that crop varieties which had performed well under previous high chemical input conditions were also performing well under the intensive organic conditions.

Institutional support for urban agriculture

The support system for urban agriculture, along with the production system it served, was seen by many in the 1990s as the potential vanguard for a nation-wide organic agricultural system. The sector was regulated and directed by the National Urban Agriculture Group (GNAU), comprising individuals from scientific and government institutions as well as urban farmers, and covering 26 sub-programmes (GNAU, 2000). Objectives and targets for these sub-programmes, which were for implementation throughout the country, are shown in Box 5.1. Urban agriculture was not only for larger cities; any conurbation from small villages upwards was obligated to have urban production of some sort.

In Havana Province, the Department of Urban Agriculture employed 292 workers, including 12–14 specialists within each of 13 municipalities. Each region had a Municipal Urban Farm Enterprise, which coordinated production, research and extension activities and networks (Ojeda, 1999). Details of the achievements of urban agriculture were regularly published in the national newspaper, *Granma*, and this publicity was used to encourage improvement. In each of the municipal people's councils was placed a qualified extensionist who liaised with farmers to develop demand-driven research programmes. Havana City had 67 such extensionists in 1998. INIFAT played a key role in coordinating urban agricultural research, which included, for example, the development of strategies for testing, evaluating and producing commercial quantities of bio-inputs (Wilson and Harris, 1996). The people's councils also

---

Box 5.1 *Objectives of the National Urban Agriculture Programme, 2001*

- To apply 10kg/m² of organic material per year to *organoponicos* and intensive gardens, and a minimum of 20t/ha on plots and patios.
- To regularly revise the sources of organic material in the municipality and at the level of people's councils.
- To create optimal conditions for the breeding of worms.
- To popularise and implement vermiculture at the level of each unit of agricultural production.
- To improve the recycling and use of urban waste.
- To link the teaching of agronomy and animal husbandry at different levels with productive urban agriculture practices.
- To achieve links with producers and each of the following: agricultural polytechnics and animal husbandry institutes, university faculties and scientific institutions.
- To raise the agro-ecological awareness of the population in environmental education while maintaining high quality production.

Source: González Novo and Merzthal, 2002

---

assisted with services such as veterinary clinics, farmers' shops, nurseries and CREEs (Companioni et al, 2002).

All inputs, such as seeds and seedlings, biological pesticides, accessories and tools, could be purchased in municipal shops (such as the *casa de semillas*, house of seeds, or *tienda del productor*, farmer's shop). These shops were staffed by qualified technicians who also provided free technical advice on the use of organic inputs. Those farm units which followed the recommendations provided could be eligible for municipal support for equipment. Specific advice on soil fertility was also provided by provincial and municipal Organic Fertilizer Reference Centres. Twelve cooperatives[3] in Havana collected and processed organic material and distributed it across the city (González Novo and Merzthal, 2002), and in 1998 a project commenced to create worm compost production centres in each province, with the eventual aim to have one in each municipality. An urban seed-savers' network was developed by the Cuban NGO – The Foundation for Nature and Humanity Antonio Núñez Jimenez (FANJ). This network aimed to conserve and use locally adapted varieties, using reliable producers to multiply seeds which were then distributed to other producers. The Foundation also produced a low-priced and widely available gardening booklet on permaculture methods entitled '*Se Puede!*' (It's Possible!) (Weaver, 1997).

A substantial amount of retraining was undertaken, of departmental staff, heads of farm enterprises, and practical on-farm training. According to one researcher, each urban producer was going on an average of four training courses a year. The knowledge base was, however, low; an evaluation of the needs of urban agriculture in 1996 (Wilson and Harris) noted a lack of understanding of the purposes behind the techniques being promoted. Urban agriculture was monitored through a set of indicators that enabled improvements in sustainability. During the 1990s, limiting factors to production were identified as a general lack of resources, technical production constraints, theft of produce, and insufficient training for non-specialist growers (Chaplowe, 1996; Wilson and Harris, 1996; Miedema and Trinks, 1998).

Impact on food security

The state maintained its investment in urban agriculture for several reasons: the high food demand in the cities, the relatively high free-market price of fresh vegetables, the need to make nutritional improvements to the basic Cuban diet, the possibility of selling direct to the consumer from the farm gate and thus overcoming postharvest losses and transport restrictions, and the potential for employment creation in urban areas (Wilson and Harris, 1996). The income from urban agriculture was crucial to supplementing the generally low state wages of the 1990s (Murphy, 1999). Ritchie (1998, p1) recounts one urban patio producer growing grapes on the roof of his building, along with vegetables and herbs in compost-filled tyres, which he sold to supplement his pension. He claimed: 'It is the duty of Cubans to find ways to support themselves, as their contribution to sustaining the gains of the Revolution.'

Urban agriculture played an important role in fulfilling dietary requirements. This could be interpreted in statistical terms, such as the extent to which it fulfilled FAO recommendations of 300g vegetables/person/day (Pagés, 1998), or, in lay terms, through ensuring that vegetables were widely available in urban zones. Urban gardens were able to supply vital vitamins and minerals, and some starches, as well as medicines and spices (Chaplowe, 1996). Foreign interest groups visiting Havana were often taken to one particular old people's retirement home, where the residents would relate that 'Our health has improved since we started working outside, our salad and fibre intake has increased, and the need for medicines has reduced.' Awareness of the wider social objectives of urban agriculture was prevalent amongst producers. The head of one private *organoponico* described their objectives as being 'To increase food supply for the local community, to increase fresh vegetable supply, and to improve the environment.' By 1999–2000, urban production was providing up to 70 per cent of daily individual vegetable requirements and about 60 per cent of all horticultural vegetables consumed in the country, according to Sinclair and Thompson (2001), or approximately 30 per cent of vegetables consumed in Havana, according to Nova et al (1999). Aside from inaccurate data sources, this variance in statistics could indicate that Havana residents had easier access to other vegetable sources, or that vegetable consumption outside Havana was much lower. Overall, urban supply was providing approximately 5 per cent of Cuba's food needs, mainly of salads rather than roots and tubers, maize or other bulky field crops.

By the end of the 1990s, evidence to the outside world indicated that Cuba was developing a unique and groundbreaking, localized agriculture and food system, at least based on reports of urban agriculture. Yet there was little information on what was going on in Cuba's rural farming systems, which produced 95 per cent of domestic food supplies. Were these also being run along localized, organic lines, and were they meeting the food needs of the population?

Notes

1   This was a relatively high salary, compared, for example, to the average earnings of a university professor of $15 or 300 Cuban pesos/month.
2   A free-living bacterium that converts atmospheric nitrogen into ammonium, making it available for plant use.
3   These cooperatives were of the new form, UBPCs, which were created in the 1990s from old state farms.

References

ACAO (1995) *Agricultura Organica*, vol 1, no 3 (December)

Altieri, M. A., Companioni, N., Canizares, K., Murphy, C., Rosset, P., Bourque, M. and Nicholls, C. I. (1999) 'The greening of the "barrios"; urban agriculture for food security in Cuba', *Agriculture and Human Values*, vol 16, pp131–140

Carrasco, A., Acker, D. and Grieshop, J. (2003) 'Absorbing the shocks: the case of food security, extension and the agricultural knowledge and information system in Havana, Cuba', *Journal of Agricultural Education and Extension*, vol 9, no 3, pp93–102

Castro, F. (1996) Speech at UN World Food Summit, 18 November

Chaplowe, S. G. (1996) 'Havana's popular gardens: sustainable urban agriculture', *WSAA Newsletter*, World Sustainable Agriculture Association, vol 5, no 22 (Fall)

Companioni, N., Hernández, Y. O. and Páiz, E. (2002) 'The growth of urban agriculture', in Funes, F., García, L., Bourque, M., Pérez, N. and Rosset, P. (eds) *Sustainable Agriculture and Resistance: Transforming Food Production in Cuba*, Oakland CA, Food First Books, pp220–236

FAO (2003) *Fertiliser Use by Crop in Cuba*, Rome, Land and Water Department Division, FAO

Funes, F. (2002) 'The organic farming movement in Cuba', in Funes, F., García, L., Bourque, M., Pérez, N. and Rosset, P. (eds) *Sustainable Agriculture and Resistance: Transforming Food Production in Cuba*, Oakland CA, Food First Books, pp1–26

Garfield, R. (1999) *The Impact of Economic Sanctions on Health and Wellbeing.* Relief and Rehabilitation Network, Network Paper 31, London, ODI

GNAU (2000) *Lineamientos para los subprogramas de la Agricultura Urbana.* Grupo Nacional de Agricultura Urbana, MINAG (Septiembre), Havana.

González Novo, M. and Merzthal, G. (2002) 'A real effort in the City of Havana: organic urban agriculture', *Urban Agriculture Magazine*, RUAF, no 6, pp26–27

HDRA (1997) *Cuba Goes Organic!* Coventry, Henry Doubleday Research Association

INIFAT (1995) *Memorias, I International Meeting on Urban Agriculture*, Havana, INIFAT

Kerry, B. R. (2001) IACR-Rothamsted, UK, personal communication (7 March)

López, F. (2000) 'El país espera por la respuesta de los orientales en el ano 2000', *Granma*, 26 January

Miedema, J. and Trinks, M. (1998) *Cuba and Ecological Agriculture. An Analysis of the Conversion of the Cuban Agriculture towards a More Sustainable Agriculture.* Unpublished MSc thesis, Wageningen Agricultural University

MINAG (1999) *Datos básicos*, Havana, Ministerio de la Agricultura

Murphy, C. (1999) *Cultivating Havana: Urban Gardens and Food Security in the Cuban Special Period.* Unpublished Masters Thesis, Fac. Latinoamericana de Ciencias Sociales, Universidad Nacional de Habana

Nova, A., González, N. and González, A. (1999) *Mercado Agropecuario Análisis de Precios.* Economic Press Service no 24, December, Havana, IPS.

Ojeda, Y. (1999) 'La granja urbana: elemento facilitador del desarrollo de la agricultura urbana', *I Fórum Tecnológico Especial de Agricutura Urbana*, Havana

Pagés, R. (1998) 'Se requiere sistematicidad y linealidad en la producción hortícola intensiva', *Granma*, 15 April

Ritchie, H. (1998) 'A revolution in urban agriculture', *Soil and Health*, vol 57, no 3 (Autumn)

Rosset, P. and Benjamin, M. (1994) *The Greening of the Revolution. Cuba's Experiment with Organic Farming*, Melbourne, Ocean Press

Roycroft-Boswell, E. (2002) 'Cuba's organic perspective', *Urban Agriculture Magazine*, RUAF, March, no 6, p25

Sinclair, M. and Thompson, M. (2001) *Cuba Going Against the Grain: Agricultural Crisis and Transformation*, Oxfam America. http://www. oxfamamerica.org/newsandpublications/publications/research_reports/art1164.html accessed May 2002

Taboulchanas, K. H. (2001) 'Oportunidades para la certificacion organica', poster presented at the *IV Conference on Organic Agriculture*, Havana

Weaver, M. (1997) 'Allotments of resistance', *Cuba Sí* (Summer), p21

Wilson, H. and Harris, P. (1996) *Report on Havana Urban Agriculture Organic Production*, Coventry, Henry Doubleday Research Association

Windisch, M. (1994) 'Cuba greens its agriculture. An interview with Luis Sanchez Almanza', *Green Left Weekly* (11 December). www.hartfordhwp.com/archives/43b/003.html

6

From Dependency to Greater Self-reliance: Transformation of the Cuban Food System

Food sufficiency is closely linked to a country's food sovereignty and its political and economic independence.

(Republica de Cuba, 1994, p71)

Directly impacted by the collapse of the Socialist Bloc, and indirectly by the changes in agricultural production, the state took a sequence of measures to fortify the Cuban food system in order to maintain the food security status of the population. These measures are discussed in terms of the three pillars of food security: food availability, food accessibility and food adequacy. Cross-cutting themes emerge, including the resilience of the Cuban food system as attempts were made to achieve food security through greater local self-reliance, the role and effectiveness of the state in this regard, the relationship between the farming approach and food security, and the degree of commonality between socialist, industrial and organic approaches to food security.

Food availability

Overall food prioritization

The socialist government had always placed a high political priority on ensuring national-level food availability. Because of this, and notwithstanding the acknowledged 'gap between expectations and results' (Nieto and Delgado, 2002), the experience and achievements of the 1990s provide an example to other countries grappling with food security problems. The 1990s saw three main national strategies to promote food security. The first was the further development of the nation's early warning monitoring system, SISVAN (the National Food and Nutrition Surveillance System). The second was the fostering of domestic food production in terms of both quantity and nutritional content. The third strategy was the aim of guaranteeing equitable access to the available food. Key to supporting these strategies was the implementation of a new National Nutritional Action Plan, which was commitment to importing necessary food and backing this up with humanitarian aid. As a result, Cuba managed to ensure and increase a basic level of food availability during the 1990s.

The political and economic context

The economic struggle of the 1990s

Cuba faced a dire economic situation in the first half of the 1990s due to the collapse of its main trading partners and the tightened sanctions of the USA.[1] In 1992 and 1993, the country experienced a strong recession, with negative GNP growth rates in excess of 10 per cent (Garfield, 1999). Neither the World Bank nor the International Monetary Fund came to Cuba's immediate aid. Because of this, Cuba had to undertake a range of economic reforms: reducing expenditures across the board, re-orientating its trade relations (mostly towards Latin America and Canada), diversifying the economy through the development of biotechnology, pharmaceutical and tourist industries,[2] decentralizing and increasing the efficiency of national institutions, attracting foreign investment through joint ventures, and allowing for the free circulation of the dollar including remittances from abroad (Alepuz Llansana, 1996).

In 1994, the decline in GNP levelled out, and government figures showed a 7.8 per cent increase in 1996. By 1999 the positive trend was continuing, with a growth rate of 4.2 per cent, and the Cuban peso, which had been operating at a dollar rate of 150:1 in 1994, was revalued at 20:1 in 1999. The budget deficit had reduced from $38.28 million in 1995 to $13.4 million in 1998 (Nieto and Delgado, 2002). Nevertheless, the Economist Intelligence Unit estimated hard currency external debt to have continually risen over this period (EIU, 1997), including outstanding arrears with the International Fund for Agricultural Development (IFAD) which stood at $14.21 million in 2000 (IFAD, 2000).

Economic policy impacts on food strategies

State economic policy directed the food strategy. Castro (1996) explained that measures were implemented to 'devote priority attention to those activities that produced export items, replaced imports or fostered sales in hard currency of exports and services on the domestic dollar market, both for the development of tourism and for the presence of foreign capital', and that measures were also taken to 'assign a priority to food production'. Certain of these measures appeared similar to the restructuring programmes of other developing countries, such as the following (Castro, 1996; Nieto and Delgado, 2002):

- continued development of different forms of organization and of mixed private and public property
- the provision of agricultural credit

- the initiation of alternative employment programmes
- development of the internal pharmaceutical and biotechnological industries.

Other measures fell into a distinctly non-mainstream context, such as:

- priority support for agriculture
- the recovery of sugarcane production
- the development of national food production to reduce food imports 'by significantly reducing the weight they represent'
- support for Territorial Food Groups, which were in charge of setting policies and making decisions over the production of food for local consumption
- the recovery of the food processing industry based on the resources available
- a reorientation of social policy to adapt to the new distribution flows of incomes of the population
- energy savings and restriction of usage.

As Nieto and Delgado note (2002, p56), the overall goal remained that of 'perfecting the socialist system, while balancing the introduction of new market mechanisms with sustained planning via foresight and anticipation of needs'. This included the 'Introduction of Perfecting the Wholesale Price System', which aimed to increase efficiency and put the agricultural economy in line with international market prices by using world market prices as the basis for internal business. Goods and services sold within the country had to reflect the real costs of imported raw materials and other production costs. Although centralized control was maintained, domestic wholesale prices were destabilized.

Easing of population growth

Easing the situation was the continuing decrease in the rate of population growth, which had been on the gradual decline for much of the previous century. Although the Cuban population had reached 11 million by 1996 (Castro, 1996), annual growth rates had reduced dramatically: from 11.1 per thousand in 1990 to 3.5 in 1995, and to 3.0 by 2000 (Pedroso and González, 1996; ONE, 2006). There was an average of two children per couple. This fall in rates was attributed to several factors: the lack of restriction on family size, accessibility of family planning, and female education and professional opportunities. Even the population in Havana City, which comprised 20 per cent of the total, was rising only gently (Zulueta et al, 1996).

Imported foodstuffs and the impact of sanctions

Food imports had previously accounted for 10–12 per cent of total imports. In the early part of the decade they rose to a 25 per cent share of the $400–500 million total. This rise came about as a result of state prioritization as well as of the 80 per cent reduction in overall import values (ODCI, 1996). Items comprised mainly grains, milk powder and cooking oil, and they were used in over three-quarters of the food consumed during that period (Casanova, 1994; FAO, 2004). Between 1994 and 1997, dwindling foreign exchange reserves meant a continued decrease in overall imports by 40 per cent, but nearer the end of the decade spending rose again to a point where food imports alone accounted for $700 million (Sinclair and Thompson, 2001).

Tighter sanctions also affected access to food at national level; sanctions added a 'virtual tax' of 30 per cent as imported goods had to come from more distant and expensive markets (Kirkpatrick, 1996; Garfield, 1999).[3] Indirect impacts of sanctions included the cancellation of favourable trade agreements and reduced access to world markets. National production was also affected: sanctions-related lost production was valued at $2 billion in 1996.[4]

Humanitarian development aid and political agendas

Humanitarian aid also played a vital role in ensuring food availability during the 1990s. In 1995, $17 million of food aid was received (Castro, 1996), and by the late 1990s humanitarian assistance (including medical aid) was valued at $1 billion (Garfield, 1999). The FAO remained active in technical cooperation, while Oxfam America supported Cuban farmers as a humanitarian response to the food crisis. The delivery of such assistance, and other forms of development cooperation, encountered problems as some sections of the international community (notably the USA) sought to undermine government institutes and instead searched out grassroots and religious groups that were less closely aligned with the state. In response to this, the state created and strengthened a raft of 'non-government organizations' in order to attract such collaboration and prevent the grassroots groups from being used for political purposes.

This political agenda continued to the end of the decade. After the Cuban drought of 1998, the USA made an offer to the appeal by the World Food Programme for emergency food relief. This offer was based on the proviso that Cuba meet certain conditions, such as allowing international monitoring of food distribution, allowing USAID markings on donated foodstuffs, and involving NGOs in the distribution process. Cuba's centralized food distribution and rationing system made these conditions difficult to comply with, and the government was also cautious of the USA's motivations in developing close links with NGOs. Political capital was made out of this impasse; a statement from a

US State Department Spokesman read 'We regret that the Cuban Government has decided to put politics ahead of the basic needs of the Cuban people' (USIA, 1998). On the other hand, at 1998 the USA was still the largest donor of humanitarian assistance to Cuba (USIA, 1998).

The National Nutritional Action Plan and the drive for increased self-sufficiency

The National Nutritional Action Plan

The crisis had curtailed the progressive moves of the late 1980s toward a National Food and Nutrition Programme (Republica de Cuba, 1994). The legacy of this gave rise, in 1994, to the formulation of a National Nutritional Action Plan, based on FAO guidelines. This Plan covered all aspects of food security and had three strategic objectives, as follows:

1   To increase and diversify national food production with the dual purpose of achieving the best possible food self-sufficiency, and obtaining foods in the most economical way and with adequate nutritional content.
2   To import those foods which are still required to complement national efforts and satisfy the demand of the population.
3   To incorporate the people as activists in this action plan, promoting self-provisioning in all regions of the country and, through this, raising food cultural awareness.

The specific objectives of the Action Plan are shown in Box 6.1.

Box 6.1 *Objectives of the National Nutritional Action Plan*

- Consolidate the UBPCs, CPAs and CCSs as the principle productive bases of Cuban agriculture.
- Improve the Agricultural Extension System.
- Achieve self-sufficiency in rice production.
- Increase the self-sufficiency of farm cooperatives.
- Combine diversified and quality food production with sustainable agriculture.
- Combine alternative agricultural practices with those of the industrialized model.
- Avoid any decrease in cultivable area per capita and attempt to incorporate saline and other marginal areas.
- Support community and family home-gardens and *organoponicos*.
- Continue the introduction of appropriate poultry production in rural areas.
- Reduce postharvest losses in the marketing chain.
- Identify and develop the potential of dryland areas according to their soils and microclimates.

Source: Republica de Cuba, 1994

This plan, approved by the State Executive Committee, was put together by a multi-sectoral ministerial group, and thus covered a broad-based range of reforms in economics, public health, agriculture, fisheries, the food industry, education, social security, and food quality and standards. It set up two levels of monitoring: at one level, each sector established its own vertical, independent monitoring mechanisms; and, at another, the Ministry of Economics established and coordinated an overall monitoring mechanism (Republica de Cuba, 1994). Annual progress reports were circulated to all the entities involved, as well as to the UN and other donors.

Plans for production increases embraced the short, medium and long term. Short-term plans involved increasing the production of bean, maize, banana, tree crops and tubers, including potato. Medium-term plans involved increasing production of the above plus wheat, citrus (including for export), vegetables, fruit, dryland crops, pork, eggs and poultry, vegetable oil crops and *organoponicos*, as well as decreasing postharvest losses. In the long term, rice, pork, eggs and poultry, milk and beef production were to be increased, with rice production aiming for self-sufficiency.

Pursuing these objectives involved a number of strategies, including the strengthening of agrarian policy through the decentralization of land and decision-making, diversification and change in land tenure; the participation of the population through community and family home gardens, distribution of small (0.5ha) plots of land, community tree planting and distribution of poultry stock; the participation of educational centres and other institutions in self-provisioning; the reduction of postharvest losses through the increase of small processing plants to conserve surplus fresh produce and the organization of an adequate labour force for harvesting; and the inclusion of nutritional considerations in agricultural development programmes. The long-term aim was to supply all nutrients from the diet rather than from vitamin supplements. However, the plan assumed that 'If the macronutrient needs are covered, that is, proteins, fats and carbohydrates, it is almost certain that vitamin and mineral needs are also achieved' (Republica de Cuba, 1994, p77).

The armed forces (the Revolutionary Armed Forces, FAR and the Youth Labour Army) were also called to help out (Lage, 1995). Murphy (1999) notes that 'At the closing speech to the army congress in 1995, the Minister of the Armed Forces affirmed that "food production is our principal task". The military would no longer take any food from civilian sources, but rather contribute by producing beyond their own food needs.'

This was in addition to new efforts made by MINAG and MINAZ to increase production through improving rural conditions, increasing salaries and providing other incentives, improving farm management, encouraging migration to rural areas, and increasing market prices and outlets, including direct marketing (Republica de Cuba, 1994; Murphy, 1999; Nieto and Delgado, 2002).

Improving rural agricultural conditions

The improvement of rural conditions had a major impact on increasing domestic food availability. Two major land tenureship changes encouraged this: the distribution of land in perpetuity, which led to a broader section of the population becoming involved in self-provisioning, and the transformation of state farms into cooperatives called Basic Units for Cooperative Production (UBPCs). By 1999, almost 3000 UBPCs had been formed, and just over two-thirds of agricultural land was held by the private sector.

Throughout the greater part of the Revolution, the state had encouraged rural to urban migration, and from 1956 to 1989 the rural population had declined from 56 per cent to 28 per cent of the total (Pérez Marín and Muñoz Baños, 1992). Reversing this process in the early part of the 1990s was not so easy. To begin with, agricultural work groups were formed to undertake production activities, mainly in plantations. In 1992, for example, these comprised 8 million workers (Oliveros Blet et al, 1998). At the same time, city dwellers were mobilized to work for 15 days in the field, and longer-term urban contingents for up to two years with the provision of attractive living conditions and pay. In 1991, for example, 146,000 residents of Havana participated in some form of agricultural work (Deere et al, 1994). New hostels were built to house all these workers. Young people were obliged to undertake agricultural rather than military service, with perks offered to them to stay on afterwards.

Meanwhile, long-term incentivization strategies were put into place, including a rise in farm gate prices of up to 50 per cent since 1997, a rise in the basic farm worker's salary (which in the late 1980s was significantly below the national average (Deere et al, 1995)), and the provision of permanent housing (Oliveros Blet et al, 1998; Pesticides Trust, 1998). Community improvement programmes included attention to housing and general community welfare, based on a new policy document 'The Dignity of Farming Life' (*La Dignidad Agropecuaria*) (MINAG, 1999). In Cienfuegos Province, for example, one ministry official explained how they were encouraging low-income groups to grow crops that could yield good economic returns. A further programme, the Plan Turquino, was offering support to repopulate and improve living conditions in marginalized mountainous zones, and to aid export-oriented coffee production (MINCIN, 2001). Through these programmes the state was addressing issues of inequality within rural areas, and as a consequence of these efforts the permanent agricultural labour force remained fairly static over the decade (although FAO statistics indicate that the agricultural labour force dropped slightly from an annual average of 868,000 in 1989–91 to 785,000 by 2000 (FAO, 2004)).

Towards the end of the 1990s a trend of urban–rural migration was occurring spontaneously as many professionals returned to family land and especially to the CCS cooperatives, largely because of the good living that could now be made out of farming. For example, in the Province of Cienfuegos, 60,000 new producers resulted between 1994 and 1997. Although the state had initially encouraged such a move, by 1999 this was felt to be no longer necessary. At 2000, according to one ministry staff member, ANAP 'would not hand out further resources for such and had no official plan to help those returning'. As a result of these efforts, as well as of increases in market efficiency, farmers and especially individuals were able, in the space of little more than a decade, to move from being a relatively underpaid section of the labour force to becoming one of the wealthiest social groups in the country (Deere, 1997).

Achievements and challenges to production increases

Main production achievements

The overriding aim of the aforementioned strategies was to increase domestic production. Figures show that this was achieved to a certain degree. According to official figures, after the low dip of 1993–94, quantities increased to a point where production for some crops has regained and even surpassed the levels achieved in 1989 (Miedema and Trinks, 1998; ONE, 1999). However, some sources, such as FAO country statistics, show a much more erratic production trend, with generally lower production quantities. Figures were skewed by the process of land redistribution that occurred during this time; figures from the newly formed UBPCs were merged with those of the longstanding *campesino* farmers – and therefore did not reflect the degree to which *campesinos* were affected by, and recovered from, the crisis, nor the performance of these new types of cooperative. In addition, figures were based on the official yields and quantities for production plans and disregarded the surpluses produced by *campesino* farmers and cooperative farms (Mesa-Lago, 1998; Miedema and Trinks, 1998). Castro (1996) also stressed that figures did not include the significant production coming from 'lots, gardens and other pieces of land'. Figure 6.1 shows the upward trend in crop production according to the Cuban National Statistics Office.

The strategies by which these increases were made varied by crop. Initially, root and tuber crops, and rice production, were prioritized by the state in order to quickly supply basic carbohydrates. Root and tuber production increased by around 80 per cent from 1995 to the end of the decade. Several factors influenced this rise: root and tuber yields had fallen by only 10 per cent during the crisis years, these crops were prioritized early on by the state (potato in particular), and the volume requirements of Acopio (the state food collection and distribution agency) made these crops attractive to the farmers in order to easily fulfil production plans. The increase in rice production was largely due to the development and growth of 'popular' or people's rice production, which took a distinctly organic approach (Funes, 2002). The growth of people's rice production is described in Box 6.2.

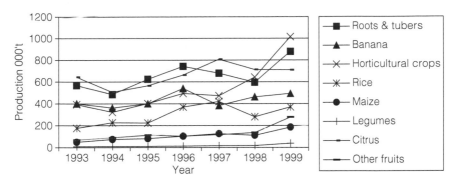

Figure 6.1 *Selected Crop Production Indicators, 1993–99*

Source: ONE, 2001

Box 6.2 *The growth of the People's Rice Programme*

The large-scale, flood-irrigated systems operated by the state were hard hit during the Special Period. Production of wet rice fell from 507,600t/year at the end of the 1980s to 147,600t in 1993. Rice was a traditional staple of the Cuban diet, especially amongst the rural population. Yet even in peak production years, Cuba had imported approximately 40 per cent of national consumption. The Nutritional Action Plan of 1994 stressed a high-input approach for a new national rice programme (Republica de Cuba, 1994), but it also introduced a People's Rice Programme, based on small farmer, low-input rice production. This Programme charged the local People's Councils with the control of production and marketing, and offered land in perpetuity for such production. In 1996, the National Rice Research Institute was commissioned to develop more appropriate, low-technology and small-scale approaches to support the Programme, and by the late 1990s approximately 40 per cent of people's rice cultivation was independent of irrigation systems. Other innovations included the identification of appropriate dryland rice varieties, the cultivation of out-of-season rice in certain areas, and the practice of transplanting seedlings rather than direct sowing. Further experimental improvements were the introduction of nitrogen-fixing trees, such as *Sesbania rostrata*, and the use of animal traction in the Asian style. Even postharvest processing was undertaken through low-technology, locally based approaches.

The aim was for local self-sufficiency. As well as this *campesino* production, state Agricultural Enterprises started to use the same low-input and small-scale approach to rice production on their self-provisioning plots. By the end of the decade, at least half of the country's rice was supplied through the people's production method. Although the small scale of production and informal, complex marketing channels make production figures difficult to measure, the approach was seen as having the potential to expand to a far wider cropping area, once external financial restrictions were lifted.

Source: Socorro et al, 2002

For other staples such as horticultural vegetables, pulses and milk, production started to show an upturn after 1995. Citrus production was greatly helped by foreign investment. Trends shows that state farms remained the major producers of citrus, and private farms of maize and legumes. State farms (which included *organoponicos*) overtook private production for horticultural crops, while private producers overtook state as the major producers of roots and tubers, bananas, rice and other fruits.

The main area where productivity remained depressed was for livestock products, as seen in Table 6.1. Most livestock products such as poultry, eggs and beef had depended on imported fodder, and production in the late 1990s was lower than a decade previously. There was little differentiation in productivity between farming type, except for a clear increase in milk production by CCS cooperatives (Statistical bulletins, MINAG, cited in Monzote et al, 2002).

Table 6.1 *Selected livestock production indicators, 1986–98*

Livestock products	Mean annual production		
	1986–90	**1991–95**	**1998**
Milk (ML)	905.9	651.7	697.7
Beef (000't)	289.1	143.1	137.3
Pork (000't)	104.5	85.5	n/a
Poultry (000't)	122.8	61.6	37.4

Source: ONE, 1997; Nieto and Delgado, 2002

Domestic production, combined with food imports, enabled calorific intake to rise from 1860kcal/person/day in 1996 to between 2200 and 2610kcal/day by 2000/01 (Nieto and Delgado, 2002; FAO, 2004). With the increase in range of food sources it became more difficult to monitor actual intake and official figures were likely to be underestimates. Still, the drive for production increases continued as a priority for the agricultural sector, in order to close the production–consumption gap. Any production increases were an achievement given that state financial subsidies had declined by 96 per cent, from 1.8bn pesos to 6.9m pesos ($9m to $0.35m equivalent), between 1993 and 1996 (Paneque, 1996).

In terms of land area, total farmed land rose from 4,060,000ha to 4,465,000ha during the period. At 1998 there had been little overall change in the proportion of land dedicated for domestic consumption as opposed to export. Sugarcane accounted for 49 per cent of permanent cropping land, and coffee and citrus for a further 6 per cent (Rodríguez, 1999). Agricultural items – raw sugar, rolled tobacco, rum, coffee, bottled honey and citrus juice – were still the main Cuban exports over, for example, biotechnology products. Fluctuations in production continued, owing to a combination of adverse climatic events, falling global prices and the continuing US economic embargo (MINCIN, 2001). Nevertheless, tobacco, coffee, cocoa and citrus production made a full recovery by the end of the decade.

Ongoing challenges to increasing production

By the middle of the decade, a number of production challenges had become apparent. These included the inefficient use of resources on state farms and UBPCs, insufficient autonomy on some UBPCs, the prohibition of certain rationed produce from entering the market (such as beef, coffee, potato, milk), the general scarcity of resources, the high delivery agreements to Acopio, tax increases, and the higher profit levels, which actually served to disincentivize some farmers from further increasing production volumes (Nova, 1995). Other causes for the continued low production in some crop sectors were the influence of the previous industrialized system and the continued loss from the food distribution system (Rosset, 1996; Wroe, 1996). Large areas of cooperative land remained in a state of neglect and the invasive *marabu* shrub had taken further hold. Lower yields in some years could be directly attributed to a combination of drought and hurricane. In October 1996, for example, Hurricane Lily struck the centre of the island, causing approximately $800 million worth of damage, including the destruction of vast tracts of banana, citrus, rice, coffee and sugarcane land (Mesa-Lago, 1998). Similarly, the outbreak of the *Thrips palmi* virus affected 45,000t of potato as well as other root crops.[6]

In 1998, an evaluation of agricultural growth in the country concluded that MINAG could create the conditions to increase production through further transformations in structure and organization, as well as through improving infertile lands, enlarging the state farm programme, transforming the payment system, linking remuneration to work results, and reinvigorating more state farms (SEDAGRI, 1998). Other mechanisms suggested for increasing production were a more supportive free market for agricultural inputs, and further reform of the ration system, specifically to introduce a dual price structure whereby lower-income households could access cheaper produce (and possibly remove upper-income households from the system completely). This would help producers by enabling them to receive a better price for their planned production, and would also encourage them to maintain their commitments to it (Deere 1997; Rodríguez, 1998).

Import, export and investment strategies

In Cuba, state control over food imports and exports was a major tool for ensuring food availability. Overall responsibility for deciding whether specific foodstuffs should be sourced from abroad or domestically rested with the Ministry of Economics, while the administration of imports and exports was the responsibility of the state Agricultural Import–Export Marketing Enterprise, CATEC. The internal tourist market was also considered to be an export market as it generated a dollar revenue, and this market was served through Acopio under the brand name 'Isla Azul', and through the tourist food procurement agency, Frutas Selectas. Private producers were not permitted to sell direct to tourists but had to trade through the relevant Agricultural Enterprise or through Frutas Selectas. Each Enterprise had different procedures concerning the passing of part of the dollar income back to the farmer.

The drive for national self-sufficiency was, for many crops, undertaken in a pragmatic way. As a general rule of thumb, exports could be undertaken when there was just about sufficient of the crop to cover basic national needs. For example, according to several ministry staff, it was easier to import rice rather than to attempt to produce all the rice nationally. Much rice was imported from China, where its production was subsidized, thus it cost less than domestic production and could guarantee a reliable and consistent supply for the ration. Domestically produced rice was channelled through the farmers' markets. Other crops, such as onion and garlic, were imported due to the relatively high costs of their production and storage in-country. Yet potato, also considerably more costly to produce and store nationally than to import, was home grown for political (paternalistic) reasons. For other crops, such as banana, there was an explicit policy to achieve self-sufficiency before developing an export market. Yet some sources suggested that this was more because the state entities involved in export were insufficiently organized to meet international quality standards rather than for any ideological reason. Regarding the potential for exporting certified organic produce, in the late 1990s it was felt to be of national importance to meet domestic food needs instead. Overall, different criteria applied to different crops, reflecting differing degrees of pragmatism, political ideology and some degree of disorganization.

In 1999, companies from at least 40 countries were investing in Cuba, the majority from Canada, Italy, Spain, France, the Netherlands and the UK. Regulations aimed to reconcile national and foreign interests, and foreign investors had to satisfy at least two of the following three conditions:

- to contribute with new technologies
- to contribute with new markets or new tourist avenues
- to contribute with financial capital and resources.

Investment in education, public health and the armed forces was not permitted, and neither was privatization or the passing of land into foreign hands (Fernández, 1998; Ferradaz, 1999). According to Ritchie (1998), the state ensured that for every dollar invested in Havana, one-third went to rebuild the city, and another third to sustain the social services.

Food accessibility

Overall strategy for food accessibility

Since the Revolution, Cuba has paid great attention to maintaining a relatively equitable society, including in terms of incomes. In many less-industrialized countries, strategies for food security centre around macro-level food availability. In Cuba as much attention was paid to accessibility at the micro level and particularly around the efficiency of distribution mechanisms to guarantee equality (Oliveros Blet et al, 1998). The ration system guaranteed a highly subsidized supply of food for the whole population, with further supplies for public institutions such as schools and hospitals (MSP, 1988; Rosset and Moore, 1997).

During the crisis, the state maintained this priority for ensuring food accessibility. In reviewing the food security situation in the late 1990s, Castro explained, 'The priority assigned to this issue, together with centralized planning, has made it possible to devote adequate attention to the objectives of satisfying the food needs of the population...' (Castro, 1996). The two strategies adopted to promote food accessibility were the diversification of food distribution channels, and the increase and diversification of sources of income available for the population. New sources of income included self-employment, hard currency remittances from relatives abroad, state labour incentive schemes and the tourist industry. The percentage of the population with access to hard currency rose from 8 per cent in 1993 to 40–50 per cent by 1996 (Castro, 1996). Despite this, there was a recognized need for assistance by those without sufficient income or hard currency to obtain certain food supplies. The mechanisms used to achieve this are described below.

Overview of the food collection and distribution system and food social security

The food distribution network

The structure of networks and flows within the food system in Cuba were controlled by the state, the main organizations responsible for this being the following:

- Acopio (Unión Nacional de Acopio, National Collection Unit), the state food procurement and distribution agency for domestic farm produce. Divided into 14 provincial Acopio Enterprises, and with over 3700 municipal sales points as well as the task of supplying the MINCIN ration stores.
- Frutas Selectas, in charge of supplying produce for tourism and related activities.

- Other national-level Agricultural Enterprises and Unions, usually commodity based, for example for seeds (Empresa de Semillas), mixed crops (Empresa de Cultivos Varios), tobacco (Empresa del Tobaco).
- Ministry for Internal Trade (MINCIN), responsible for administering the ration system through its provincial, municipal and local networks. The ration stores channelled both imported and locally produced (and processed) foodstuffs, through La Bodega (the warehouse – for food products) and Las Placitas (the agricultural sales points – for fruit and vegetables). These ration stores also sold non-rationed, subsidized items. MINCIN also operated a nationwide network of wholesale and retail stores and gastronomic centres.

Within this structure, the National Nutritional Action Plan specified four types of destination for agricultural produce (other than self-provisioning): the ration organized by MINCIN; social feeding such as to hospitals and schools; industry and processing; and tourism. In addition, domestically produced goods also found their way onto the black market.

The ration system and social security

The Social Security System (which comprised assistance for the most vulnerable, specifically old people, disabled people, single mothers, children and youth) and the ration system had been established in 1979 and 1962 respectively, and provided a solid food security structure during the challenging years of the early 1990s. During this period, this equitable form of food distribution and controlled sale, or *canasta basica*, was the only market that the state was able to maintain. An increasing number of food products became rationed: 19 rationed items in the 1980s increased to virtually all food items in the early 1990s (Murphy, 1999). Its composition also changed: animal protein content was reduced and vegetable protein and high-calorie foods increased (Felipe, 1995).

During this period, the state managed to maintain its purchase contracts with farmers so as to guarantee the availability of rationed products. At least 80 per cent of the main crops produced were contracted in this way (Rodríguez, 1998). Agricultural products would be collected by Acopio at the farm, or delivered by the farmer to the local Acopio centre, and then transported to centralized urban markets from where the goods would be transferred to neighbourhood distribution centres which were specialized for crops, meat or dairy (Enríquez, 1994).

Orientation towards provincial and local self-reliance

At the World Food Summit in 1996, Castro stated, 'Food production here [in the country's four mountain ranges], as well as for the rest of the country, is approached from the point of view of viable crops to attain self-reliance, with the aim of cutting down the food supplement they require.' This illustrated the shift in emphasis from a centralized distribution system to one of partial self-sufficiency and localized production–consumption chains at provincial level. Each Provincial Agricultural Delegation was responsible, through Acopio, for meeting local demand as far as possible with local supply. Direct sales (*tiro directo*) and the farmers' markets were the two main mechanisms for achieving this, as well as the increase in on-farm and household self-provisioning as described in Box 6.3.

Box 6.3 *The growth of farm and household self-provisioning*

The setting aside of land for self-provisioning on all farms was encouraged. On state farms, UBPCs and CPAs, the goal was to bring unused land into production for annual and perennial food crops and livestock 'modules', to feed members. These were tended communally. CCSs already had a self-provisioning strategy for owner-members and in the early 1990s this was extended through the provision of individual plots for farm workers. These individual plots served as a huge incentive and helped CCSs to retain labour, and the same strategy was then adopted for individuals on state farms (Deere et al, 1994). Other individuals could also be provided with land if they could demonstrate a justifiable cause, although the state continued to ensure that collective production outweighed individual. Land parcels of up to 0.5ha – too small for incorporation into cooperatives – were distributed to retired persons or others outside the production sector interested in self-provisioning, with the opportunity to sell any surplus produce. By 1998, perpetuity rights for almost 11,000ha of land had been given to 45,800 people or *parceleros* (Enríquez, 2000). This also diminished dependency on the ration system.

Source: Wright, 2005

The organizational structure for supplying the ration was extensive. The Ministries of Domestic Trade and of Public Health, along with the Institute for Internal Demand, drew up the provincial production plans, based on quantities and nutritional balance needs. The Institute for Internal Demand determined how much of each crop would go to each destination. This was based on calculations of food availability which were undertaken on a monthly basis at municipal level. Acopio was then responsible for sourcing produce from other provinces to fill any deficit. Generally, Acopio buyers advised farmers as to which markets they might best sell at. However, during periods of scarcity, certain crops had to be sold to Acopio and not elsewhere, and if a province could not supply its own food needs, it looked to neighbouring provinces to assist. For example, during one period in 1999, farmers in one region of Havana Province received a circular from MINAG which instructed them to direct all their banana, potato and sweetpotato crops to Acopio, this including the surplus that they had planned to sell at the higher-priced farmers' markets. These

instructions were followed, farmers accepting that they would not get the premium price expected. In another instance, the demand by Acopio for bananas
was such that, in one municipality of Havana Province at least, farmers were
unable to retain sufficient to meet their own needs. Some crops did not fit into
this provincial planning. Because of its importance in the Cuban ration, potato
distribution continued to be dealt with at national level. This led to situations
where, for example, a province was experiencing a shortfall in supply of roots
and tubers for its ration but was still transporting potatoes to Havana. Surpluses
destined for the processing industry also crossed provincial borders: for example, seasonal surpluses of mangoes in Cienfuegos could be sent to a processing
plant in Ciego de Avila Province.

Debate over ways to improve provincial and municipal self-sufficiency were
held on an ongoing basis. At one such meeting in Cienfuegos, in 1998, topics
under discussion were concerned with the means to increase efficiency, to
reclaim degraded lands, to organize cottage industry for food conservation, and
to prepare for climatic and other disturbances (SEDAGRI, 1998).

Agricultural professionals tended to agree that complete self-sufficiency at
a provincial scale was not appropriate or possible, owing to the differing agro-
climatic conditions of each and to the comparative advantages of growing
particular crops. Maize and bean were well suited to eastern provinces such as
Holguin, for example. Nevertheless, there was also agreement that basic staple
crops could be produced at the local level, and most provinces were already
self-sufficient in certain crops. Cienfuegos, for example, was seasonally self-
sufficient in vegetables and fruits, and was working to increase sufficiency in
lowland rice.

The Food and Nutrition Surveillance System (SISVAN)

Shortfalls in supply continued to be identified and addressed by the Food and
Nutrition Surveillance System, SISVAN, which was responsible for food availability, accessibility and adequacy. SISVAN was coordinated through the
provincial network of the Institute for Food Hygiene and Nutrition (INHA).
The mandate of INHA's provincial centres was to bring together the provincial
government with representatives from the agricultural, commercial, educational
and public health sectors. At quarterly meetings, a nutrition specialist would
present knowledge of regional needs, based on data from the regular nutritional
surveys undertaken by the public health teams of the Institute for Internal
Demand. The agricultural representative would then present production figures
(actual and forecasted), and strategies would be adopted for making good any
shortfall, either from a neighbouring province or by importing through the
Ministry of Exterior Trade. In the east of the country, where food security problems were more severe,[7] these meetings were held more frequently, and when
there was insufficient national purchasing power to satisfy needs, the World

Food Programme assisted with food donations. This surveillance system included an early warning system (Alerta Acción) to identify when interventions might be needed and to organize advance back-up. By the end of the decade, a major intention of SISVAN was to create a reserve of basic food supply in order to cover short-term emergencies.

SISVAN continued to be considered the most advanced surveillance system of its kind in Latin America and its staff continued to consult and provide training overseas. Gay et al (1986) identified the main reason for its success as being its broad horizontal and vertical reach, coordinating and unifying the full range of specialized and sectoral institutions involved with food issues, at national, provincial and municipal levels.

Improving the efficiency of state food collection

Up until 1995, Acopio was a highly centralized collection, distribution and marketing organization, purchasing agricultural produce from both state and private farms for distribution through various government channels. In that year, Acopio was merged with the Mixed Crops Enterprise to form a marketing, production and service provision entity. This merger was partly in order to increase efficiency – Acopio became less centralized and could use the food collection and distribution points of the Enterprise, but the merger was also a response to the formation of the UBPCs in that same year. The UBPCs took over responsibility for approximately half the land that the Mixed Crops Enterprise had previously been managing, and thus a large proportion of its activities passed into private hands. After the merger, the Mixed Crops Enterprise retained its installations: storehouses, machinery workshops and irrigation equipment and so on, providing these as a service to farmers. Despite the merger, the Enterprise and Acopio were still often thought of, and referred to, as separate institutions. Acopio continued to establish contracts with farmers, specifying annual production plans and monthly delivery plans (*plan de entregar*). Failure to sell could lead to confiscation of land (Enríquez, 2000). Prices paid were relatively low – farmers complained that they scarcely covered production costs – but were guaranteed.

For most of the 1990s, fuel shortages and storage of produce were major problems. In an analysis of rice and grain conservation in Havana City, Méndez Núñez et al (1998) identified a long list of storage problems, including leaking roofs, lack of ventilation and lack of knowledge by quality controllers. These problems instigated change. In 1993, a new procurement process was introduced, focusing on the creation of multiple rural collection points (Puntas de Acopio). *Campesinos* would bring their produce to a predetermined point at a particular day and time, where they would hand over to Acopio and receive payment on the spot. In Havana, the number of Acopio staff was increased to deal with this. There were some teething troubles, as Deere et al (1994) remarked:

'The probability of an Acopio representative, a truck and a *campesino* all converging at the same point at the same time was highly unlikely.' In the early 1990s, much of the produce collected was passed through regional wholesale hubs (*concentrados*), which were set up to receive, improve – by grading, cleaning and so on – and distribute produce to large urban conglomerations. (These centres subsequently changed to focus only on agricultural processing.) Alternatively, produce could be transported straight from the farms to the sales points in whatever Acopio vehicles still existed, and this direct delivery cut down warehouse storage losses by more than 10 per cent. At the same time, where financial resources allowed, refrigerated warehouses were constructed to conserve agricultural produce, mainly potato, which the state guaranteed to supply all year round.

With the fragmenting of large state farms and the resultant increase in the number of farm units, Acopio had to increase its transport network at local levels. In one municipality of Havana Province, for example, the number of lorries had increased from 14 to 25. Operations were also being undertaken on a more timely basis; around Havana City for example, Acopio was collecting perishable salad crops very early in the morning to reach Havana in time for the agricultural markets the same day. This saved on storage costs. The impact of fuel shortages on food transport around the country is discussed in Box 6.4.

Box 6.4 *The transport connection*

Fuel shortages were a major driving force for a more regional approach to food accessibility. These shortages stimulated a resurgence in rail cargo, at least for those products which had been historically transported in this way, such as sugar, honey, petrol, minerals and coal (Alepuz Llansana, 1996). However, much investment would be required to improve the basic rail infrastructure and facilities, and over short distances road freight was more competitive (Ruíz González, 1996), with trains more economically beneficial over longer distances (Henríquez Menoyo, 1996). Efforts were made to increase the efficiency of road transport through standardizing transport schedules, improving information sharing between storehouses and suppliers, and upgrading containers and other equipment. The merger between Acopio and the Mixed Crops Enterprise was intended to improve transport of products to the market, but Ruiz González (1998) observed that this merger did not result in the hoped-for results, because of poor vehicle maintenance and poor hand-over of statistical and accounting mechanisms.

Source: Wright, 2005

Changes to the marketing system over the 1990s

The early reinstatement of the parallel market

Acopio had to become more organizationally efficient; it was now chiefly responsible for providing a stable food supply, including the coverage of any periods of shortfall and with as little recourse to imports as possible. This meant increasing the capacity of the domestic processing industry. As Castro (1996) explained: 'Producers are being contracted to cooperate with processing plants to ensure a steady quantity of raw material throughout the year.' The 'parallel market' of domestically processed foodstuffs, which had given way under the crisis, was reinstated as early as 1992, with outlets selling preserved meat products, dairy and conserved fruit and vegetables. Unlike food from the ration system, these products were not subsidized and so prices were much higher, but they were in more abundant supply.

The short-lived direct sales mechanism, tiro directo

The introduction of a direct sales mechanism, *tiro directo* ('direct throw'), was a major move towards a more localized production–distribution system, where supply agreements were made between a specific cooperative and a regional state market, without the involvement of Acopio but based on capped prices. It was of particular benefit for perishable products, which had previously often languished for several days before being collected by Acopio. Transport was variably covered by the cooperative or by the market. According to Enríquez (2000), *tiro directo* was very popular with producers, who were rapidly coming to see their cooperatives as business enterprises and calculated the cost benefits of delivering their produce against waiting for it to be collected or having to guard it against theft. Enríquez (2000, p7) cites one analyst, who predicted that 'with the sub-division of state farms into UBPCs and the overall trend toward smaller-scale production, farmers would be increasingly encouraged to do this because Acopio would be unable to attend to the growing numbers of producers, each of whom has a limited amount of produce'. In some regions, *tiro directo* was the main supply channel. However, the state was not willing to encourage a more widespread usage of this strategy, and towards the end of the 1990s attempted to replace it with other mechanisms. Perhaps, as Enríquez suggests, this was because the state did not want to lose its control of the capped price markets that it operated through Acopio.

Partial liberalization of agricultural markets

In order to provide an official source of food surplus to the ration, the state created a range of subsidized, municipal agricultural markets. One of the most common was termed Las Placitas, which supplied fresh produce at subsidized prices, and this was part of the state retail network or *commercio minorista*. Sales prices in these markets remained capped but were variably higher than those available through the ration store La Bodega. Produce for these markets was sourced from the regional wholesale hubs, through Acopio, or directly from the farmers themselves. The emergence of urban agriculture as another new supply source made centralized food distribution far more difficult to organize. Acopio offices found it difficult to deal with this increase in complexity and went through a difficult period in the mid-1990s.

Yet these changes were still not sufficient to resolve the shortfall in food supply in the early part of the decade. Even with the rapid state measures to provide more food outlets, food prices continued to rise, even in the ration system. Instances arose where *campesinos* were refusing to accept Cuban pesos for their produce and instead were demanding dollars or scarce consumer goods as payment (Deere et al, 1994). Not only had the production sector declined in the early part of the decade, but also the food industry had practically ceased to exist, and with it the networks of workers' cafeterias. There was just not enough food available to be purchased; at the most critical time, in 1993, people held on average 15 months' salary in hand with nothing in the shops to buy (Nieto and Delgado, 2002).

Inevitably, under such conditions, the black market thrived, offering food but at extremely high prices. It sold the excess from the *campesino* sector together with any produce deemed of too poor a quality to sell to Acopio, and, during 1990–93, the incidence of thefts from fields and warehouses rose to supply this market. The main products available were meat and dairy, grains, and roots and tubers. One hundredweight of garlic, which Acopio would purchase for 130 Cuban pesos ($6.5), would sell on the black market for 1000 Cuban pesos ($50) (Enríquez, 1994). To reduce the hold of black marketeers, in mid-1994 the state permitted circulation of the US dollar and allowed access to the dollar shops by the Cuban population.

Creation of 'free' farmers' markets and private food businesses

In order to further reduce the power of the black market, and also to stimulate production and increase efficiency, to complement the ration, and to increase access to fresh produce, legislation was passed in September 1994 to permit 'free' farmers' markets for the sale of food surplus to the ration. Within two weeks, 121 such markets had been established all over the country, organized by MINCIN. The speed of execution showed what was possible when political will aligned with farmers' needs, and also gave an indication of the latent capacity and entrepreneurship of farmers in responding to better market opportunities.

Prices were self-regulating to reflect supply and demand, a mechanism that had not previously been permitted. This resulted in increased competition between producers and sellers and contributed to a slow reduction in food prices. Some price regulation was maintained to prevent a recurrence of the problems experienced with the previous *campesino* farmers' markets of the 1980s.[8] The state collected a sales tax and levied charges for the services that the market provided. Indirect control was achieved by analysing product prices on a monthly basis, and flooding the capped price markets with a similar but cheaper product if prices of particular crops were seen to be rising too high. Prices varied between these farmers' markets, being some 20 per cent higher in Havana City than in the provinces (Nova, 1995). This was partly the result of lower taxes in the city, which incentivized farmers to make the longer journey to the capital. Lower taxes were also applied at markets where there was a need to enhance supply (Murphy, 1999).

These markets came to play a crucial role in enhancing food access, by making excess goods available to more people at regular times, locations and prices. This also benefited the producers, who in turn reduced their black market activities. The option of selling surplus at a higher price contributed to farmers doubling or tripling their production (Sinclair and Thompson, 2001). Some differentiation between crop quality was also visible. For example, Acopio purchased maize by quantity, whereas in the farmers' markets the most important factor for consumers was the size of the cob. Another feature of the markets was the sale of some processed foodstuffs, ranging from sweets and drinks to meals, the prices of which fell between 1995 and 1997 (Deere, 1997).

Farmers were thus provided with a range of obligatory and non-obligatory outlets for their produce. They were obliged to satisfy their own needs, those of the state plan, their social responsibilities to hospitals or nurseries, and then the needs of export markets, tourism, industry and seed supply. All these commitments vied with supplying the farmers' market, and market officials checked that farmers had first met their production quotas for Acopio before permitting sales. One farm extensionist explained this: 'In the future we could get a place on the farmers' market, but not this year. We have and want to meet the national demand first – through Acopio and then Frutas Selectas.' These commitments aside, anyone who worked the land could sell there, either directly or through a sales representative. This raised another problem: the elimination of middle men. The state wanted to avoid or reduce speculation on basic food products, and for this reason was experimenting with concentrating the marketing in the hands of a few officially designated sales representatives (Oliveros Blet et al, 1998). Access to these markets was further aided by the decentralization of the transport structure, this enabling the free contracting of vehicles. Three years into their establishment, it was the state farms and private CCS and individual producers who were taking most advantage of the farmers' markets. UBPCs and CPAs were supplying only 10 per cent of produce, as shown in Table 6.2. A contemporary evaluation put the contribution of individual farmers at a rate of 78 per cent of sales values (Deere, 1997). Around the mid-1990s, the farmers' markets were channelling nearly all the grains and fruits (noncitrus) and were also a major contributor to the supply of roots and tubers. Still, unlike the lower priced alternatives, the farmers' market did not guarantee sales, and Acopio continued to purchase between 50 and 80 per cent of total farm production (Oliveros Blet et al, 1998).

Table 6.2 *Participation in the supply of crops to the farmers' markets, by farm type*

Farm type	1994	1995	1996	1997
State	24%	21%	36%	41%
UBPC	15%	7%	6%	5%
CPA	19%	8%	4%	4%
Private	42%	64%	54%	50%

Source: Nova González, 2000

Another new option was permission for the private sector to sell prepared foods. Up to that time, the state had taken this role, but the majority of its establishments had disappeared. By August 1995, 36,864 licences had been granted for private food businesses (*Granma*, 1995). Again, a substantial degree of control and regulation was implemented, with, for example, a maximum limit on the size of privately run restaurants and high tax rates. Control levels fluctuated as the state attempted to prevent inequitable individual gain. In 1996, for example, new taxes drove down the number of legally registered self-employed workers from 208,000 to 180,000 (ODCI, 1996).

The resulting range of marketing options for farmers

At 2000, five price levels were being paid to farmers by Acopio, in ascending order: the lowest price paid was for food destined for the ration and social feeding; then state food processing (the parallel market); then the capped-price agricultural markets; then the farmers' markets where prices were regulated by supply and demand; and finally the highest purchase prices were for food destined for the domestic, tourist and export markets. In practice, the sales channels varied by crop but also by quality.

Taking bananas as an example, Figure 6.2 sets out the marketing channels for banana in the Province of Havana at 2000. Because banana yields in Havana Province were high and of adequate quality to supply the tourist market, so bananas were marketed largely through the Mixed Crops division of Acopio, which dealt with domestic tourist demand. For the ration, any variety of banana was used. Bananas in the farmers' market sold for $0.3–0.35/kg, which was considered a high price but one that consumers were willing to pay when the fruit was ripe and could be eaten immediately. Lower-quality fruits went for 1.5c each ($0.15–0.2/kg), and those which had been artificially matured for 2c each. The tourist markets were supplied with both the more traditional Cavendish banana variety and the modern FHIA varieties. Cavendish had shown to be more popular with tourists because of its longer shelf life and thus good appearance. It was also more popular with farmers owing to its stronger fragrance,

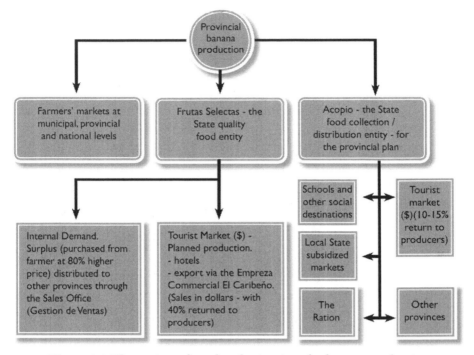

Figure 6.2 *The variety of market destinations for banana production in Havana Province*

Source: Wright, 2005

which masked the ripening chemicals used. There was a limited demand by hotels for 'healthy' bananas (*platano sano*).

Trends and challenges in the food system at national and local levels, at the end of the 1990s

Subsequent trends and challenges to the food system

The overall impact of all these changes was dramatic, according to one of the few contemporary accounts available, that of Oliveros Blet et al (1998) at the Faculty of Geography of the University of Havana. Food supply increased, and the influence of the black market was diminished, with black market prices falling by some 40 per cent. Urban agriculture also made a significant impact by providing competition to the farmers' markets so that prices in these markets decreased by 30 per cent between 1994 and 1997 (Deere, 1997). Although urban agriculture supplied only 5 per cent of the volume of total marketed produce, this produce constituted a nutritionally important part of the Cuban diet that had previously all but disappeared from the market. It included a variety of vegetables and salads, fruits and some grains. Overall, food prices remained fairly stable during the second half of the decade, dropping only slightly between 1996 and 1998 (Nova et al, 1999).

Yet, towards the end of the decade, prices in the farmers' markets of MINCIN were still often out of reach of a large sector of the population. Prices of produce from urban *parceleros* were some 20 per cent lower, as were the capped markets of MINAG. This difference was most extreme in Havana; in 1999 there were 2383 state-controlled markets and 332 farmers' markets throughout the country, but in the capital this ratio was reversed, with 70 farmers' markets and only 13 under state control. In the first quarter of 1999, for example, sweetpotato was priced in comparative units of 1.11 in the farmers' market compared to 0.47 in the capped price markets. For banana (fruit) the difference was 1.53 to 0.28, for onion 4.63 to 2.66, and for rice 4.43 to 3.31 (Recio and Jiménez, 1999). In an analysis of the relative price immobility in the farmers' markets, Nova González (2000) identified that supply was constricted by the high profits and organized cartel-type approaches of the private farming sector.

Poorer segments of the population were also unable to access the dollar stores, which were, by the end of the 1990s, the only suppliers of some basic staple products outside the ration, such as cooking oil. The black market was still important, although its dominance and the extreme pricing had reduced. Inequality continued, although Deere (1997) suggests that inequality would grow faster without farmers' markets.

Staff of Acopio, who had struggled with the expanding range of produce sources and end destinations in the middle of the decade, were by 2000 feeling that they had weathered the most difficult period and could now better appreciate the value of diversification into a wider range of food suppliers. There were suggestions that Acopio could even decentralize further, if the CPA and CCS cooperatives would sell directly to Las Placitas. One municipality in Havana Province, for example, had 23 such Placitas, and eight sales points, all which were currently managed by Acopio in terms of ensuring supply. Sales taxes had been one constraint for *campesinos* selling at these markets, but the state had halved this tax (from 15 per cent to 7.5 per cent). Yet in the City of Havana, the complexity of sales networks was such that the logistics of distribution might still need supervision and could not, it was felt, be decentralized.

In the late 1990s, computerized systems were being introduced to manage the ration at provincial levels. More importantly, however, the role of the ration was diminishing significantly. In 1996, around 40 per cent of national food needs were supplied through this means, and the ration in Havana City was supplying approximately 61 per cent of daily calorific requirements, 36 per cent of animal protein, 65 per cent of vegetable protein and 38 per cent of fats required (Nova González, 2000). By the end of the decade, only 10 per cent of food needs were being supplied by the ration. In one municipality of Cienfuegos

Province, for example, Acopio was channelling only potato through the ration; the other crops it collected and distributed through Las Placitas. In fact, the state was aiming to eliminate the ration book completely, although it recognized that many families still depended on it until food supplies could be stabilized. In the late 1990s, the Social Security System was benefiting over 1.5 million people, or 14 per cent of the population.

To replace the ration, planned produce was being increasingly channelled through capped markets of the state retail network and the agricultural sales points. At the end of the 1990s, new types of sales points were being developed, with Las Placitas being incorporated into a system of 'new-style market places': Placitas del Nuevo Tipo and Placitas Precio Tope. Prices at these markets were approximately 20 per cent lower than in the farmers' market, depending on the season.

There were also challenges to be overcome in terms of the farmers' marketing of products, with many cooperatives still in need of training, including on how to access local sales points. Training was generally less available in more remote regions of the country, where there was poorer access to transport.

Supply to the domestic tourist market was also impacted. Frutas Selectas had for many years been a loss-making subsidiary of Acopio. Because of this, and the rise in tourism, in 2001 responsibility for its management was transferred to the Citrus Corporation, which had a long history of successful export production. Frutas Selectas continued to purchase up to 80 per cent of farmers' produce for tourist destinations. Although prices paid to farmers were in Cuban pesos, they were higher than Acopio's and came with the provision of imported inputs and tools. Frutas Selectas also worked to stimulate supply to meet the tourist demand by, for example, cooperating with MINAG to produce out-of-season fruits. At the end of the decade, farmers were becoming aware of the need to supply a higher-quality product for this destination.

The restructuring of the Cuban economy therefore required further adjustment to improve food accessibility, an ongoing process, which Castro (1996) explained: 'The changes comprising, on the one hand, the introduction of market mechanisms and the adjustment of their instruments to our objective reality and, on the other hand, staying within the sphere of planning for anticipation and forecasting, will continue to be aimed at improving socialism.'

The resultant food system at national and municipal levels
The institutional relationships of the food system are displayed in Figure 6.3, showing the range of food supplies available to the consumer, some of which were provided almost free, and others which required purchasing. Although food imports and aid played a role, domestic production formed the main foundation of the food system.

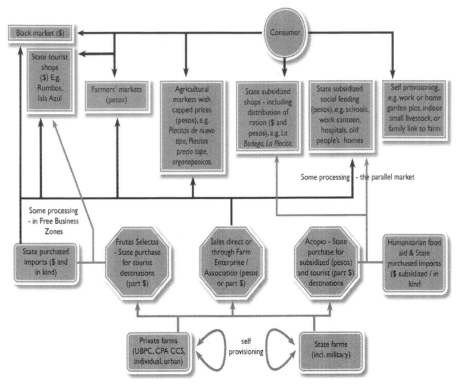

Figure 6.3 *The institutional relationships of food system linkages in Cuba, at 2000*

Source: Wright, 2005

At the micro level, a useful analysis of the organization of food supply in one municipality of Havana was made by Palet Rabaza and Piedra Castro (1998). This analysis attempted to differentiate and evaluate the relationship between locally managed components and more centralized ones. A summarized translation of this analysis is given in Box 6.5. It identifies the full range of institutions and actors involved in the food system, as well as the complementary and positive interactions between a semi-decentralized state and the private sector.

Box 6.5 *The organization of food supply in one municipality of Havana City*

Plaza de la Revolucíon is one of 15 municipalities in Havana City. Covering 12.3km², it is subdivided into eight administrative units, or People's Councils. Although the official number of inhabitants is 164,000, unofficially this number is much larger. Being the location for 18 ministries and approximately 500 institutions, enterprises and companies, it has a working population of 97,000 which it has to feed during the working day. In addition, it has to support

16 hospitals, 116 schools with over 12,000 pupils, the University of Havana and five other higher educational centres.

The food chain within the municipality may be classified as those organizations that are under local management and those that are not. Those with less local management or participation play a greater role in guaranteeing basic foodstuffs for the population. These and their roles are as follows:

- Provincial wholesale enterprises – which decide on the supply strategy (quantity per capita, frequency of delivery, transport and so on) according to product availability.
- Production enterprises with a national range – specifically of roast coffee, meat, poultry, dairy, drinks and liqueurs and fish, each of which has its own methods of direct distribution to the retail network.
- Municipal enterprises of marketing and collection of agricultural products – intermediaries for the control and organization of distribution to the population.
- Establishments of the commercial retail network – direct contact points with consumers, consisting of agricultural sales points (Las Placitas), supermarkets, warehouses (la Bodega), dairy, and fishmongers and butchers.
- School canteens – social feeding in schools is undertaken in a centralized way through the Provincial Food Production Enterprise.

Those components of the food chain with a higher level of local management (though not necessarily private) are more recently established, and comprise the following:

- Gastronomic networks – decentralized and outside the official supply system, this network includes restaurants, cafeterias, nightclubs, ice cream parlours and Basic Supply Units. It uses mainly hired private transport.
- Basic Supply Units (UBAs) – created to reduce prices in the private sector, these units facilitate the buying, selling and manufacturing of food products, and also organize a network of mobile (bicycle) food points, again in competition with the private entities.
- Workers' canteens – almost all work places have these. As it was found to be difficult to maintain these based on traditional state food suppliers, so their management has been put in the private hands of each workplace.
- Self-provisioning areas – related to the workers' canteens. In the Municipality of Plaza productive land is scarce, so only 18 of the enterprises and ministries have been able to develop self-provisioning areas. These are located towards the outskirts of the city, with surplus produce going to schools, nurseries and retirement homes.
- Farmers' markets – Plaza contains only three of the capital's 49 markets, but these turn over 15 per cent of the total volume, and include on-site UBAs that produce meals using ingredients from the market.

In addition, three new components are also under local decentralized management:

- Commercial and gastronomic networks emerging from the economic recovery – these were created with foreign investment and consist of hotel chains and commercial businesses operating restaurants, cafeterias, snack bars and shops. They are only accessible to dollar customers but supply some essential products to the private sector.
- Private-sector gastronomic outlets – self-employed, legal individual and family initiatives that both create employment and increase food availability, especially for the local workforce. The main forms are privately run restaurants (*paladares*), and sales from home or from a fixed sales point.
- Urban agriculture – though restricted by land availability in Plaza, 79 horticulturalists receive support from the People's Councils, and nine *organoponicos* are in operation. Products are destined firstly for workers' canteens, then sales to workers and other local centres, and then direct sales to the population.

Strict centralization of the retail network has been necessary to ensure equity. Nevertheless, decentralization of certain components has improved management, even though these components lack financial and legislative resources. Farmers' markets and UBAs are good examples of local management, and the expanding tourist service sector is achieving its objectives of obtaining foreign exchange and creating local employment. Meanwhile, the rapid rise of private-sector gastronomic outlets provides an example of the capacity of local people to take the initiative even without formal organization or proper resources. Even with the increasing decentralization and local management, local government and People's Councils will continue to play an important role in facilitating the diverse range of actors involved in the local food system.

Source: Palet Rabaza and Piedra Castro, 1998

Household food security coping strategies

For the household, prices of many food markets and outlets fluctuated, as did the availability of products due to seasonal variations and to the ability of the state to import. Daily newspapers paid much attention to the current state of food supply in the agricultural markets (Recio and Jiménez, 1999).[9] Meanwhile, the quantity of food available through the ration system was decreasing.

To ensure food sufficiency in this situation, households devised various coping strategies, depending on their situation and capacity. The following three boxes give some idea of how three particular and contrasting households organized their food arrangements in 2000. Box 6.6 describes the strategies of a rural farming family, which is a member of a CCS cooperative in the east of the country. The second, Box 6.7, details the approach of a provincial urban family. Box 6.8 contrasts these with the coping strategies of an urban extended family living in the capital Havana.

Box 6.6 *Hardworking but constrained: food security strategies of José, a CCS farmer in rural eastern Cuba*

José comes from a farming family; he, his father and brother worked on a state farm up to 1993 and since then on family land as part of a CCS. José was trained as a veterinarian, and has since undergone two training courses on farm management: at the national ANAP training school and locally with the Mixed Crops Enterprise.

On starting a family of his own, he applied for and was given 24ha for himself, to expand his dairy operations through a state programme to increase school milk supplies. Though he describes the land as good quality, it was covered in *marabu*, a hardy, thorny invasive shrub, which he had to clear by hand. Fortunately, he had access to an old tractor constructed of spare tractor pieces, which he shares with his extended family. José also had to install electricity, paying a sum of two pigs in-kind to the state electricity company to lay approximately 3km of new cable. He would like more land, but the municipal MINAG office has refused his request so far. He sees other people obtaining more land, through bribery, but he refuses to do this as he is saving his capital to invest in stocking the land. José had been fortunate in obtaining capital to purchase 11 cows through his previous involvement in the tourist industry.

José's CCS cooperative is linked to the local Livestock Enterprise. The cooperative would like to apply for more support or 'strengthening' by the state, but it does not yet meet the basic eligibility requirements – it lacks a proper tractor and basic tools. As a result, members cannot officially purchase subsidized inputs but instead have to pay black market prices, which are two to three times higher. Although José had plenty of experience with industrial agriculture in the late 1980s, he feels that the main constraint on productivity is the lack of tools and machinery to undertake the basic land tasks, rather than the lack of chemical inputs. His extended family help each other out through exchanging seed, labour, produce and so on.

José has to sell some of his produce – cheese, milk and meat – to the state. This includes all the beef he produces; home consumption of beef is illegal. Not being 'strengthened' limits the market opportunities of the CCS, especially as he lives approximately 7km from the nearest town and a three-hour bumpy bus ride from the provincial capital. The taxes he would have to pay at the farmers' markets are prohibitive. His main income comes from selling approximately 90 per cent of his cheese on the black market, to women who then bus it up to Havana.

José receives a typical state ration. This comprises the following basic items, per month: 1lb dark sugar (0.4c/lb), 6lb white sugar (0.75c/lb), 5lb rice (6c/lb), 6oz beans, 6oz pulses, 2lb tinned ham. He can also buy roots, tubers, fruits and other products from the state ration store at higher, but still subsidized, prices. At this store, 1lb of rice costs 17.5c. José receives a special diet for his baby daughter including, sometimes, beef and milk. So, along with the ration, José produces his own food, largely pork, milk, cheese, wheat and horticultural crops. He also exchanges produce with family members. He has access to other peso markets including the agricultural sales points and the farmers' market, although prices there are less affordable for him, and more variable. Sugar, oil, more rice and fat, all have to be bought in the dollar shop or on the black market. For a young, ambitious and hardworking farmer such as José, the weak access to land and markets are limiting his potential productivity.

Source: Wright, 2005

Box 6.7 *Resourceful with remittances: food security strategies of the Conejo family living in a provincial town*

The Conejo family own a four-roomed, one-storey house on the outskirts of town. Their youngest son lives at home and brings in $7.5/month from his job in a local processing factory. Sra Conejo describes in detail their food ration per month, as follows:

Per person: 5lb rice (1c/lb), 0.5 litre oil, 2–3lb salt, 6lb white sugar (0.8c/lb), 1lb brown sugar (0.4c/lb), one bread roll/day (0.15c/unit), ¼lb beans or pulses, 1lb fruit banana (2.5c/lb), 6oz soy meat, 1–2lb roots and tubers, one tin fish to share between three people (7–10c/tin), 3–4 eggs (0.75c/unit), ½ bar soap, one toothpaste to share between three people, 12oz coffee, 4oz tea, 3 litres cooking fuel, 2 gallons petrol. Other items, such as fruit and meat, may be included but their quantity and availability varies each month. Old people and children are entitled to extra rations including 1litre milk/day and cereals.

Sra Conejo recalls when the ration provided for all their needs, including pork that was sold at 5c/lb. Nowadays they can purchase pork from the state market at $0.75/lb, or from the agricultural market at $1/lb, and fat at $0.75/lb. For vegetables they go to the *organoponico*, and for extra oil, soap and detergent to the dollar shop. Black market street hawkers always have something to offer, such as pork (at $0.75/lb) and cheese. These sources are cheaper than the farmers' market because no tax is paid, and they also have a more reliable supply. Sra Conejo has heard that people in Havana receive more through their ration including fish, and one whole bar of soap per month.

The family keeps two pigs and five chickens in the 2m^2 backyard of their house. Sr Conejo gets up early in the cool morning to prepare sweets using sugar he saves from the ration and from other sources, and garden fruits that he can obtain almost free. He produces about 70 sweets a day, and sells these unofficially via a middleman. This brings in 2.5c/unit, which amounts to $1.75/day minus his minimal production costs. The Conejo family are fortunate: one son lives abroad and sends home dollar remittances. With these they can purchase some luxury items and are planning an extension to their house. They consider themselves resourceful; with low financial overheads, they can live a relatively comfortable life.

Despite all the changes, including substantial positive improvements, there remained a widespread feeling of food insecurity in society. On farms, even those producing abundantly, this insecurity focused around the lack of specific types of food, in particular of meat and other primary proteins. Off-farm insecurity could be partly attributed to the still-precarious sources of food or finances to access it. Dependency on a daily state ration, with a fluctuating composition, meant that households were unable to stock up with food reserves – a problem compounded by the lack of cool storage facilities in many households – and did not feel in secure control of their food future.

Box 6.8 *Feeling insecure: food security strategies of the Sanchez family in Havana City*

Sra Sanchez is the matriarch of an extended family of seven, living in a three-bedroomed apartment in a prosperous central suburb of Havana. Her son, his wife and their two teenage children sleep in one bedroom, with Sra Sanchez, her daughter and other granddaughter in another. The third bedroom is left empty. Her son owns a flat in the neighbouring suburb that he chooses to rent out. With all three grandchildren studying, and Sra Sanchez and her daughter-in-law minding the house, the only wage-earners are her son and daughter, who both hold full-time jobs – although they are able to take substantial time off whenever help is required.

Prior to the Revolution, Sra Sanchez was a nightclub singer and entrepreneur, and the change in government had brought an end to the glamorous life upon which she and her husband had thrived. In the 1960s, and with two young children, they decided to move to Spain. Her husband went ahead – but shortly afterwards state emigration laws changed and Sra Sanchez and her children were unable to follow. Lacking practical skills, she had survived by informal wheeling and dealing. For this family, the major income generator is their property. Their colonial apartment is grand and spacious, and two rooms have been converted into a separate apartment that they rent out to foreign visitors. Together with the rent from the son's house, they have a substantial dollar income that they use to purchase imported material goods, and to upgrade and enlarge their house further.

With these dollars the family are also able to supplement their ration with sufficient amounts of meat, special-occasion luxuries and other additions, which they purchase through the black market or from local state stores. There is always a queue at these stores, but Sra Sanchez or her daughter-in-law have the time to wait. Agricultural sales points (La Placitas) at both ends of their block sell a limited range of produce directly from farms in the Province. On an average day, these points might be selling bananas, oranges and sweetpotato. The farmers' market is 15 minutes' walk away. These markets, however, are rarely visited by this family. They choose to eat few fresh vegetables other than lettuce and tomato. They fry most of their food and consume quantities of processed, sweet foods and drinks. They feel food insecure, constantly discussing food and their insufficient access to it. Given their adequate financial resources, the deficiency may be more one of knowledge or will to improve dietary habits. The perceived insecurity is not only about food; the family also has a wall-to-ceiling cupboard bursting full of medicines and pills, which they accumulate from their foreign guests.

Food adequacy

Policy on food adequacy in relation to agriculture

In Cuba, adherence to the socialist principle of meeting the human right to adequate food has resulted to some degree in recognition of the link between agricultural production and adequate diet. This recognition was evident, for example, in the general objective of agriculture in the National Nutritional Action Plan: 'To increase agricultural production, aiming to completely satisfy the nutritional needs of the population, based on increasing efficiency of use of human, material and financial resources, whilst at the same time conserving the environment' (Republica de Cuba, 1994, p60). The plan set out specific objectives for ensuring food adequacy in terms of nutritional composition, diversity and freshness, as follows:

- Develop farmers' markets and self-provisioning plots to provide fresh foodstuffs.
- Increase the supply of fresh and processed grains, roots, vegetables and fruits to provide dietary fibre and a varied diet.
- Increase the supply of animal protein-rich foods: eggs, meat and milk.
- Increase the supply of vegetables oils.
- Increase sales of poultry chicks for home-fattening.
- Increase planting of fruit and timber trees on farms.

The plan set out specific strategies to meet its aim of improving health and nutrition levels, as follows:

- Raising awareness in schools and in the adult of the risks of an inappropriate diet.
- Reducing the quantities of refined salt and sugar used in agro-processing.
- Promoting the consumption of fresh vegetables, roots, grains and fruits in both social and individual feeding.
- Creating centres for micronutrient nutritional vigilance in all provinces.
- Increasing availability of foods rich in vitamins A and C.

The plan also gave consideration to meeting international ISO food standards and to setting up a network of laboratories for the certification of foodstuffs and water. It noted that SISVAN paid attention to the nutritional adequacy of social food provisioning, such as the ration. However, in practice, and aside from urban agriculture, the prevailing drive to increase production yields overshadowed health concerns within the agricultural sector. Little attention was paid to micro-level health issues and their relation with agriculture. Instead, the issue of human health was dealt with separately by the Ministry of Public Health.

Institutional responsibilities and post-crisis changes in health research

The Institute for Nutrition and Food Hygiene (INHA) in Havana, under the Ministry of Public Health, was a key contributor on food security issues, as well as being responsible for managing SISVAN. The Institute was also responsible for identifying and dealing with both the undernourished and the obese. A further responsibility was developing nutritional guidelines for the National Nutritional Action Plan – following WHO guidelines – and then developing dietary recommendations based on these. At 2000, these recommendations were not yet complete.

In the 1990s, the Institute had become somewhat independent from the ministry and had doubled in size to 169 employees, of which 56 were researchers. The constraints of the Special Period had impacted public-funded research on food and nutrition, and the Institute was limited to whatever could be undertaken in national currency. The lack of resources affected both research topics and publications. López Espinosa and Díaz del Campo (1996) describe how the principal food journal, the *Revista Cubana de Alimentacion y Nutricion*, continued to publish throughout the early 1990s, but less frequently and with far fewer articles: an average of 16.5 articles per issue as opposed to 23 previously.

Health fluctuations and trends in the 1990s

Overall decline in nutritional levels

According to FAO figures, the number of undernourished people in Cuba increased by 11 per cent between the periods 1990–92 and 1997–99. This was partly the result of cutbacks in state food provision: for example, schools had previously provided children up to the age of 13 with milk, but after the crisis this age limit dropped to six. (Children under the age of five maintained similarly high nutritional indicators as in previous years.) Other health indices showed a similar continued decline: the incidence of low birth weights rose from 7.3 per cent to 8.8 per cent between 1992 and 1996. The ongoing sanctions were also culpable: during the most critical period, annual deaths directly attributable to the sanctions was 7500 adults over the age of 65, and this figure was the lowest estimate (Garfield, 1999; Sinclair and Thompson, 2001).

Overall, the decline in nutrition generally was greatest during the first part of the decade, after which the rate of undernourishment fell and obesity rose. To put this in perspective, the health of vulnerable groups in Cuba still remained one of the best in Latin America (Sinclair and Thompson, 2001).

Changes in dietary habits and the return of 'Western' disease

The basic Cuban *campesino* diet comprised rice, beans, roots and tubers, meat and some salad. Regional variations existed: less processed food and more traditional roots, milk, meat and rum were consumed to the west of the island; more fried roots and tubers in the centre; and more rice and soups ('slave food') to the east (Nuñez Gonzalez and Gonzalez Noriega, 1995). As a response to the crisis, people were encouraged to consume vegetable-based foods instead of animal protein. Since 1994, food scientists started to substitute the loss of dairy with soya products, particularly flavoured yoghurts for children. Over 40 dairy factories were converted to process soya, which also produced soya-based cheese, ice cream and chorizo (Montanaro, 2000). From 1991, Cuba started to import vegetable oils, transforming the basic source of dietary fat from animal to vegetable. The incidence of heart disease fell by almost half between the late 1980s and 2000: from 160 per 100,000 to 88 per 100,000. Its lowest rate, in the early 1990s, corresponded with the decline in pork fat at that time. Yet because of the continued consumption of low-grade, saturated fats, heart disease remained the main cause of death. Sugar consumption increased dramatically as a proportion of the daily diet, owing to the lack of alternatives, As recently as 1999–2001, sugar and its products composed 21 per cent of total dietary energy supply (FAO, 2004).

At the end of the 1990s, the availability of certain foodstuffs remained very low – particularly marine fish, which was a result of the combination of over-fishing and demand from the growing tourist market. Cubans mostly had to rely on canned and freshwater fish. For others foods, availability had increased, such as of vegetables introduced and made popular by the Chinese community that were then taken up by urban gardeners. The promotion of urban agriculture meant that some segments of the population were eating more healthily. One ministry official explained: 'The dietary habits of the population have changed a lot. Vegetables used to be called weeds, but now they are accepted in the diet. Vegetables used to be the most expensive and unobtainable part of the diet, but now they are available – urban agriculture supplies half of the city's vegetable needs.' Yet vegetable consumption remained low; this was attributed to the high costs of the farmers' markets, and also to the lack of tradition and culture – many households chose to purchase processed, imported foodstuffs (as in the example in Box 6.8). Similarly, unprocessed wholefood products, such as brown rice and wholemeal bread, found little mainstream acceptance among consumers, even though these could be purchased cheaply in one of a small chain of peso health shops. Brown bread in particular was felt to perish more rapidly in the tropical climate if not packaged properly.

Unsurprisingly, the number of obese and overweight people mirrored the dietary trends over the ten-year period: in Havana City, 30 per cent of the population was overweight or obese in the 1980s; this dropped to 16 per cent in 1993, but rose again to 36.5 per cent by 1999 and was continuing to rise as food availability increased. At the end of the decade, the national monitoring of obesity was reinstated by INHA, and the results of this would be used to educate the population on nutritional guidelines, via radio, television and other media. In this sense, the three- to four-year food crisis was long enough to reduce some Western diseases, but not long enough to change habits. By 2000, the most common causes of death amongst the population were the same as those in the

1980s (in order of frequency): heart attacks, cancer, cerebro-vascular, accidents, pneumonia, suicide and diabetes. Levels of osteoporosis had doubled, which may be due to the decrease in milk consumption (although much milk was available on the black market).

There was a clear need for more awareness-raising on nutrition. Early on it was apparent that there was a lot of convincing to be done to encourage Cubans to change their dietary habits. Education and active participation were identified as the key to solve this problem (Windisch, 1994). Yet public participation and active involvement in public health programmes remained low, with most Cubans adopting a passive approach in their use of the public health service (Garcia Caraballo, 1996). To counteract this, the Ministry of Health commenced support for the development of methodologies for resolving community food problems at the level of People's Councils. Other positive moves included those of the Committee for the Prevention of Disease Through Dietary Modifications, based at the National Cancer Institute in Havana, which ran education programmes on the link between diet and disease, and promoted vegetarian diets, organic foods and exercise (Montanaro, 2000). One nutritionist researcher suggested that the population had been spoiled by the state's attempt to provide for their unhealthy food preferences. Although appearing passive receivers, 'The people's way of resisting is to refuse to eat what may be offered to them, and so the state has learned not to offer what may be rejected.' In this way, the state had arguably opted for an indulgent rather than a benevolent form of paternalism, at least where food was concerned.

Several challenges impeded the improvement of dietary quality. By necessity, the state ration focused on bulky, energy-dense items such as starchy carbohydrates over more nutrient-dense foods. Individuals could only receive their ration from the local food distribution point at which they were registered, and this limited access to foods for people who needed to travel at short notice. The quality of products was low, because of both the lack of farm-gate quality standards and the siphoning-off of better farm produce to the more profitable tourist market. In fact, there was a clear and increasing differentiation between the quality of food available to the Cuban population compared to tourists. This quality included the amount of chemicals applied, and also the potential use of genetic engineering. For example, maize in Havana City could be sold to tourist hotels and restaurants through Frutas Selectas, which selected only first-grade maize and was increasingly demanding chemical-free products. As tourism was becoming more important to the Cuban economy, the service industry was becoming more professional and in particular targeting the up-market, wealthy tourists. The restaurants and official private eating establishments (*paladares*) reflected this with relatively high prices. Peso food outlets remain poor quality. Such differentiation strengthened the view amongst the population of the luxury lifestyle in the outside world.

Trends in food safety and health concerns

INHA was responsible for monitoring food product contamination, such as levels of mercury in fish, nitrates, lead in vegetables produced near roads, and aflotoxins in cereals. Prior to the crisis, INHA had been examining the issue of residue levels in food crops (mainly nitrites in horticultural crops) destined for export to Eastern Europe. This research was dropped after exports stopped, and the need to feed the people took priority over eliminating contaminants. Because of a lack of resources and especially reagents, research in the 1990s tended to focus on only the most vulnerable crops and on biological rather than chemical contamination. The incidence of food-related disease outbreaks fell rapidly in 1992, due to the changes in sources and forms of food consumption. It then rose from 1993 onwards because of the increased consumption of locally caught fish and of street foods, especially sweets and meats. Street foods were particularly prone to carrying disease, because of the level of disrepair of refrigeration units and the poor food habits of food preparers and vendors (Grillo Rodríguez et al, 1996). The privatization and decentralization of agricultural production also made it more difficult to monitor and control food safety.

Studies of agricultural produce between 1990 and 1995 demonstrated generally low levels of pesticide and fungicide residues. There were major exceptions to this, and particularly of EBDC (ethylenebisdithiocarbamate) fungicides and other pesticide residues in tomato, onion and potato, which exceeded maximum permissible levels (Dieksmeirs, 1995; Vega Bolanos et al, 1997). These particular crops were prioritized in terms of agrochemical input application. Research demonstrating the toxic effects of pesticides led the Ministry of Public Health to first regulate and then prohibit the use of organochlorines (in 1990 and 1995). Overall, the number of individuals reported as suffering from pesticide toxicity increased up to 1992 and from thereafter decreased to 1995, although figures were likely to be underestimates as overall incidence of toxicity was under-registered. Mortality rates due to toxicity during this time were low, at just under two per 100,000. Generally within the agricultural sector there were low levels of concern over the toxic impact of agrochemicals.

The link between agricultural production and human nutrition was little evident. INHA and INIFAT were working to evaluate the impact of an increase in vegetable consumption on one target community. There was no obvious work on the impact of change to lower-input agriculture during the 1990s on dietary health, and health professionals had a low awareness of organic agriculture and its relation to food quality and lacked the resources to undertake research on such. In one nevertheless interesting case, Rodríguez-Ojea et al (1998) made a link between mountainous rural populations and a deficiency in iodine. Although this deficiency was seasonal and thought to be caused by the fluctuating access to imported foodstuffs, a correlation was found between iodine deficiency and levels of soil erosion. The authors suggested that the promotion of food self-sufficiency had actually induced a less varied diet, one that was lacking the previous diversity of imported foods. This study led to the creation of a multi-sectoral National Committee for the Control and Eradication of Iodine Deficiency Disorders in Cuba, as well as a nationwide monitoring system for the control of iodized salt production and distribution. This incident highlighted two issues: that the state approach to nutritional deficiency was one of direct substitution rather than looking to the cause of the problem (in this case a soil deficiency)[10]; and that the promotion of food self-sufficiency required measures to ensure that local natural resources were capable of supplying sufficient nutrient requirements.

During the 1990s, national programmes were established by the Ministries of Food, Industry and Health, to counteract major macronutrient deficiencies – of iron, vitamin A, and calcium. These included fortifying children's food, and, in 1998, importing flour supplemented with iron. Although food supplements were acknowledged as short-term measures, health professionals were concerned that providing nutrients through direct food sources would necessitate changing the balance of crops being produced and changing food habits, both of which would require longer-term awareness raising.

Notes

1   To compound the Torricelli Act of 1992, the Helms-Burton Act of 1996 set punitive measures against third-country companies investing in property in Cuba that was previously owned by US citizens (Giles, 1997).
2   By 1999, tourism provided 50 per cent of Cuba's foreign income (Figueroa, 1999).
3   However, sources at the Bureau of Inter-American Affairs of the US Department of State put this figure much lower at only 2–3 per cent (USIA, 1998).
4   Garfield (1999) explains the deeper, more devastating impact of sanctions: 'Trade embargoes cause macroeconomic shocks and economic and social disruption on a scale that cannot be mitigated by humanitarian aid, and which affects the well-being of a population well beyond their state of health.'
5   During the latter part of the decade, a number of Cuban 'NGOs' strengthened their intention towards achieving food security, including the Cuban Animal Production Association (ACPA), the National Association of Small Farmers (ANAP), the Cuban Association of Agricultural and Forestry Technicians (ACTAF), the Cuban Council of Churches (CIC), and the Cuban Association of Sugarcane Technicians (ATAC) (Nieto and Delgado, 2002). These NGOs had the advantage of providing more attractive collaboration opportunities for foreign donors.
6   *Thrips palmi* is a pest of the Cucurbitaceae and Solanaceae families. Its outbreak in Cuba was perceived as an external factor, emanating from the USA. Cuba filed a complaint against Washington, under the UN International Convention on Biological Weapons, for introducing this pest that had never before existed in the country. It

was verified that a US crop-dusting plane did fly over Cuba on the date in question. (For more detail, see Zilinskas, 1999.)

7   According to staff from the Ministry of Health, the east of the country was harder hit by the decrease in food availability because production in this region was more geared towards the production of export cash crops, and this slowed down the move to increase self-provisioning.

8   In 1994, Deere et al (p229) had suggested that 'It is doubtful, however, whether the defamed Free Peasant Markets of the 1980–86 period will be re-opened.' The re-opening of farmers' markets in the 1990s indicated the character of the Cuban government, in revisiting the free market system that had previously given such poor results, and also reflected the political will for experimentation and pragmatism, even over ideological considerations.

9   A wide range of channels were used to disseminate information on food and agriculture in the country. These included through ANAP, the People's Councils, the Federation of Cuban Women, the Young Communist Union and the state-run media: television, radio and newspapers. MINAG maintained a press and public relations group that was responsible for obtaining an overview of public opinion. There was a high level of public awareness on agricultural issues, and continuous dissemination of news on the state of national production, such as on the sugarcane harvest, the quality of cigars, and any prevalent pest or disease outbreak.

10   Soil iodine levels may be increased by the addition of seaweed fertilizers. Alternatively, iodine-rich crops may be grown or supplied, such as garlic, lima beans, spinach or squash. A lack of zinc in the diet may act as a suppressant, as can the consumption of foods such as maize and cassava.

References

Alepuz Llansana, M. (1996) 'El transporte y la economía cubana de los noventa: acciones ante a la crisis', in Chias, L. and Pavón, M. (eds) *Transporte y Abasto Alimentario en las Ciudades Latinoamericanas*, Cd. Universitaria, Mexico, Universidad Nacional Autónoma de México, pp315–326

Casanova, A. (1994) 'La economía de Cuba en 1993 y perspectivas para 1994', *Boletín informativo, CIEM*, no 16 (July), pp3–14

Castro, F. (1996) Speech at UN World Food Summit, 18 November

Deere, C. D. (1997) 'Reforming Cuban agriculture', *Development and Change*, vol 28 (1997), pp649–669

Deere, C. D., Pérez, N. and Gonzales, E. (1994) 'The view from below: Cuban agriculture in the "Special Period in Peacetime"', *The Journal of Peasant Studies*, vol 21, no 2 (January), pp194–234

Deere, C. D., González, E., Pérez, N. and Rodríguez, G. (1995) 'Household incomes in Cuban agriculture: a comparison of the state, co-operative and peasant sectors', *Development and Change*, vol 26, no 2 (April), pp209–234

Dieksmeirs, G. (1995) *Plaguicidas Residuos: Efectors y presencia en el medioambiente.* Unpublished doctoral thesis, Havana, INHA

EIU (1997) *Country Report: Cuba. 4th quarter 1997*, London, The Economist Intelligence Unit

Enríquez, L. J. (1994) *The Question of Food Security in Cuban Socialism*, Berkeley, University of California

Enríquez, L. J. (2000) *Cuba's New Agricultural Revolution. The Transformation of Food Crop Production in Contemporary Cuba.* Development Report No. 14, Dept. Sociology, University of California. www.foodfirst.org/ pubs/devreps/dr14.html

FAO (2004) *Food and Agriculture Indicators*, Cuba. www.fao.org/countryprofiles

Felipe, E. (1995) 'Apuntes sobre el desarrollo social en Cuba', *Boletin informativo, CIEM*, no 20 (March–April), pp3–16

Fernández, P. (1998) *Cuba Toward the Third Millenium.* Seminar by Pablo Fernández of MINAG Cuba, 9 October, Wageningen University

Ferradaz, I. (1999) 'Ponencia del Ibrahim Ferradaz Garcia. Ministerio para la Inversión Extranjera y la Colaboración Económica de Cuba', *Revista Bimestre Cubana de la Sociedad Económica de Amigos del País*, vol 85 (January–June), epoca 3, no 10, pp64–75

Figueroa, V. (1999) *Revolución Agraria y Desarrollo Rural en Cuba.* Paper presented at conference on co-operativism, October, Santa Clara, UCLV

Funes, F. (2002) 'The organic farming movement in Cuba', in Funes, F., García, L., Bourque, M., Pérez, N. and Rosset, P. (eds) *Sustainable Agriculture and Resistance: Transforming Food Production in Cuba*, Oakland Ca, Food First Books, pp1–26

Garcia Caraballo, C. (1996) *Seguridad alimentaria participacion en una communidad del municipio de Matanzas.* Tesis de maestria, Havana, INHA

Garfield, R. (1999) *The Impact of Economic Sanctions on Health and Wellbeing.* Relief and Rehabilitation Network, Network Paper 31, London, ODI

Gay, J., Grillo, M., Castro, A. and Plasencia, D. (1986) 'Sistema de vigilancia alimentaria y nutricional en Cuba: desarrollo y perspectives', in Gay, J. (ed) *Memorias del Taller Internacional sobre Vigilancia Alimentaria y Nutricional*, INHA, Havana, pp4–21

Giles, S. (1997) 'Cuban deals', *Health Service Journal*, no 107 (27 November), pp26–29

Granma (1995) 'Resumen del encuentro de Presidentes de las Asembleas Municipales del Poder Popular', *Periodico Granma*, 21 September

Grillo Rodríguez, M., Lengomín Fernández, M. E., Caballero Torres, A., Castro Domínguez, A. and Hernández Alvarez, A. M. (1996) 'Analisis de las enfermedades transmitidas por los alimentos en Cuba', *Revista Cubana de Alimentación y Nutrición*, vol 10, no 2, pp100–104

Henriquez Menoyo, E. (1996) 'El transporte para el abasto de ciudad de La Habana', in Chias, L. and Pavón, M. (eds) *Transporte y Abasto Alimentario en las Ciudades Latinoamericanas*, Cd. Universitaria, Mexico, Universidad Autónoma de México, pp339–360

IFAD (2000) *Project Portfolio Performance Report* (July), Rome, PL Division, IFAD

Kirkpatrick, A. F. (1996) 'Role of the USA in shortage of food and medicine in Cuba', *Lancet*, vol 348, pp1489–1491

Lage, C. (1995) 'La economía cubana en 1994', *Boletín informativo CIEM*, no 19, pp187–200

López Espinosa, J. A. and Díaz del Campo, S. (1996) 'The first 8 years of the Cuban publication of food and nutrition', *Revista Cubana de Alimentación y Nutrición*, vol 10, no 1 (January–June). http://www.bvs.sld.cu/revistas/ ali/vol10_1_96/ali01196.htm

Méndez Núnez, G., Acosta Pérez, B. and Fraga, R. (1998) 'Análisis de la conservación del arroz y los granos en su comercialización en Ciudad de La Habana', in Interián Pérez, S., Henríquez Menoyo, E. and Chías Becerril, L. (eds) *Seguridad del Abasto Alimentario en Cuba y México: producción y logística*, La Habana, Editorial Grupo IT, pp173–190

Mesa-Lago, C. (1998) 'Assessing economic and social performance in the Cuban transition of the 1990s', *World Development*, vol 26, no 5, pp857–876

Miedema, J. and Trinks, M. (1998) *Cuba and Ecological Agriculture. An Analysis of the Conversion of the Cuban Agriculture towards a More Sustainable Agriculture.* Unpublished MSc thesis, Wageningen Agricultural University

MINAG (1999) *La Dignidad Agropecuaria*, Holguin, MINAG

MINCIN (2001) Ministry of Domestic Trade, Cuba. www.cubagob.cu/Ingles/des_eco/mincin/mincin.htm

Montanaro, P. (2000) *Cuba's Green Path: An Overview of Cuba's Environmental Policy and Programs and the Potential for Involvement of US NGOs*, California, Cuba Program, Global Exchange. www.globalexchange.org

Monzote, M., Muñoz, E. and Funes Monzote, F. (2002) 'The integration of crops and livestock', in Funes, F., García, L., Bourque, M., Pérez, N. and Rosset, P. (eds) *Sustainable Agriculture and Resistance: Transforming Food Production in Cuba*, Oakland Ca, Food First Books, pp190–211

MSP (1988) *Situación nutricional del país.* Informe Técnico, Havana, Ministerio de Salud Pùblica, Havana

Murphy, C. (1999) *Cultivating Havana: Urban Gardens and Food Security in the Cuban Special Period.* Unpublished masters thesis, Fac. Latinoamericana de Ciencias Sociales, Universidad Nacional de Habana

Nieto, M. and Delgado, R. (2002) 'Cuban agriculture and food security', in Funes, F., García, L., Bourque, M., Pérez, N. and Rosset, P. (eds) *Sustainable Agriculture and Resistance: Transforming Food Production in Cuba*, Oakland Ca, Food First Books, pp40–56

Nova, A. (1995) 'Mercado agropecuario: factores que limitan la oferta', *Revista Cuba-Investigación Económia*, INIE, no 3 (October)

Nova, A., González, N. and González, A. (1999) *Mercado Agropecuario Análisis de Precios.* Economic Press Service No. 24 (December), Havana, IPS

Nova González, A. (2000) 'El mercado agropecuario', in Burchardt, H. (ed) *La Ultima Reforma del Siglo*, Caracas, Nuevo Sociedad, pp143–150

Nuñez Gonzalez, N. and Gonzalez Noriega, E. (1995) 'Deficiencias regionales en las comidas tradicionales de la populacion rural de Cuba', *Revista Cubana de Alimentación y Nutrición*, vol 9, no 2, pp79–93

ODCI (1996) *Cuba Factbook*: 1–7. www.odci.gov/cia/publications/factbook/cu.html

Oliveros Blet, A., Herrera Sorzano, A. and Montiel Rodríguez, S. (1998) *El Abasto Alimentario en Cuba y Sus Mecanismos de Funcionamiento.* Unpublished manuscript, Faculty of Geography, University of Havana

ONE (1997) *Estadisticas agropecuarias. Indicadores sociales y demográficos de Cuba*, Havana, Oficina Nacional de Estadísticas

ONE (1999) *Anuario Estadístico de Cuba 1999. Edicion 2000*, Havana, Oficina Nacional de Estadísticas

ONE (2001) *Anuario Estadístico de Cuba 2001*, Havana, Oficina Nacional de Estadísticas

ONE (2006) *Anuario Estadistico de Cuba 2006*, Havana, Oficina Nacional de Estadísticas
Palet Rabaza, M. and Piedra Castro, A. M. (1998). 'Abasto alimentario y gestión local: caso
del municipio Plaza de la Revolución en Ciudad de la Habana', in Interián Pérez, S.,
Henríquez Menoyo, E. and Chías Becerril, L. (eds) *Seguridad del Abasto Alimentario
en Cuba y México: producción y logística*, La Habana, Editorial Grupo IT, pp201–214
Paneque, A. B. (1996) 'Agricultural reforms: are they working?', *Granma International* (27
November), pp8–9
Pérez Marín, E. and Muñoz Baños, E. (1992) 'Agricultura y alimentación en Cuba',
Agrociencia, serie Socioeconómica, vol 3, no 2 (May–August), pp15–46
Pesticides Trust (1998) 'Cuba's Organic Revolution', *Eco-Notes*. www.ru.org/81econot.html
Recio, R. and Jiménez, E. (1999) 'Subieron o bajaron los precios?' *Trabajadores*, 14 June,
p 9
Republica de Cuba (1994) *Plan Nacional de Accion Para la Nutrition*, Havana
Ritchie, H. (1998) 'A revolution in urban agriculture', *Soils and Health*, vol 57, no 3,
Autumn
Rodríguez, S. (1998) 'La comercialización de la producción agrícola', *La Transformaciones
de la agricultura cubana a partir de 1993*. Proyecto de Investigacion II/71/972,
Capitulo V. Havana.
Rodríguez, S. (1999) *Cuba: la Evolución y Transformación del Sector Agropecuario en los
Anos Noventa*. Economics Press Service, no 14 (July), Havana, IPS
Rodríguez-Ojea, A., Menéndez, R., Terry, B., Vega, L., Abreu, Y. and Díaz, Z. (1998) 'Low
levels of urinary iodine excretion in school children of rural areas in Cuba', *European
Journal of Clinical Nutrition*, vol 52, pp372–375
Rosset, P. M. (1996) 'Cuba: alternative agriculture during crisis', in Thrupp, L. A. (ed) *New
Partnerships for Sustainable Agriculture*, Washington DC, World Resources Institute,
pp64–74
Rosset, P. and Moore, M. (1997) 'Food security and local production of biopesticides in
Cuba', *Leisa Magazine*, vol 13, no 4 (December), pp18–19
Ruiz González, L. A. (1996) 'Análisis comparativo del transporte automotor y ferroviario
en el abasto a cortas distancias', in Chias, L. and Pavón, M. (eds) *Transporte y Abasto
Alimentario en las Ciudades Latinoamericanas*, Cd. Universitaria, Mexico, Universidad
Nacional Autónoma de México, pp327–338
Ruiz González, L. A. (1998) 'Proposiciones para organizar el transporte y la distribución
de productos provenientes del agro en Ciudad de La Habana', in Interián Pérez, S.,
Henríquez Menoyo, E. and Chías Becerril, L. (eds) *Seguridad del Abasto Alimentario
en Cuba y México: producción y logística*, La Habana, Editorial Grupo IT, pp125–132
SEDAGRI (1998) *Actualidades de la Agricultura*, Boletin 0/98. Ciudad de la Habana,
Agencia de Informacion Para la Agricultura (AGRINFOR)
Sinclair, M. and Thompson, M. (2001) *Cuba Going Against the Grain: Agricultural Crisis
and Transformation*, Oxfam America. http://www. oxfamamerica.org/newsandpubli-
cations/publications/research_reports/art1164.html accessed May 2002
Socorro, M., Alemán, L. and Sánchez, S. (2002) '"Cultivo Popular": small-scale rice pro-
duction', in Funes, F., García, L., Bourque, M., Pérez, N. and Rosset, P. (eds)
Sustainable Agriculture and Resistance: Transforming Food Production in Cuba,
Oakland Ca, Food First Books, pp237–245
USIA (1998) *US regrets Cuban rejection of emergency food relief* (1 October).
www.usia.gov/regional/ar/us-cuba

Vega Bolanos, L. O., Arias Verdes, J. A., Conill Diaz, T. and González Valiente, M. L. (1997) *Uso de Plagicidas en Cuba, su Repercusion en el Ambiente y la Salud*, Havana, INHA

Windisch, M. (1994) 'Cuba greens its agriculture. An interview with Luis Sanchez Almanza', *Green Left Weekly*, 11 December 1994. www.hartfordhwp.com/archives/43b/003.html

Wright, J. (2005) *Falta Petroleo. Perspectives on the Emergence of a More Ecological Farming and Food System in Post-Crisis Cuba*. Doctoral thesis, Wageningen University

Wroe, A. (1996) 'Oranges and lemons: the limits of farm reform. Heroic illusions. A survey of Cuba', *The Economist* (6 April), pp8–10

Zilinskas, R. A. (1999) 'Cuban allegations of biological warefare by the United States: assessing the evidence', *Critical Reviews in Microbiology*, vol 25, no 3, pp173–227

Zulueta, T., Pedrosa, D. and González Rego, R. (1996) 'Proyecciones de población para la ciudad de La Habana 1990–2010. Metodología y resultados', in Chias, L. and Pavón, M. (eds) *Transporte y Abasto Alimentario en las Ciudades Latinoamericanas*, Cd. Universitaria, Mexico, Universidad Nacional Autónoma de México, pp59–75

7

Cuban Food Production in the 1990s: A Patchwork of Approaches

With the Special Period, the state soon realized that Cuba had to become more food self-sufficient in order to substitute for the previously imported foods and inputs. This meant increasing efficiency of the agricultural sector in two ways: through the streamlining and improving of production techniques and technologies as well as of farm organization; and through providing incentives for increased production. Farmers across the country thus had to develop coping strategies to deal with both the challenges of the Special Period and these new state reforms in the agricultural sector. What emerged was a patchwork pattern of farming, with divergent forms of agriculture existing on-farm, and between farms, regions and provinces. This mixture of approaches was still in a state of flux towards the end of the 1990s.

State efforts to increase agricultural efficiency in the 1990s

State reforms in production techniques and technologies

The status of imported farm inputs in the 1990s
In the mid-1990s, world crude oil prices rose, and this, coupled with disputes with Russia over sugar deliveries, contributed to ongoing supply fluctuations in Cuba's agricultural sector and continued serious fuel shortages. According to Cuban data sources, oil imports were at a low of 1.2Mt in 1995 and from then on rose slightly to 1.6Mt by 2000 (ONE, 2006). However, US data sources suggest far higher figures, even if derivatives are included, with imports fluctuating between 7.6 and 8.4Mt oil per year between 1993 and 2000 (EIA, 2008). Both sources concur that domestic oil production more than doubled over the same period, reaching between 2.1 and 2.7Mt production in 2000. Over the same period, electricity generation rose by approximately 20 per cent.

Agrochemical input prices and quantities rose over the decade but remained lower than in the 1980s. Pesticide imports (which had dropped from 20,000t in 1990 to a low of just under 5,000t in 1993) rose steadily to almost 10,000t by 1998 (CNSV, 2000). Between 1999 and 2001, fertilizer imports rose

30 per cent from 96,000t to 137,000t (although consumption was still only 20 per cent of the 536,000t imported in 1989) (FAO, 2003). The national production of fertilizers was still not back on its feet by the end of the 1990s, with plants in Cienfuegos and Matanzas remaining idle. A consequence of the fuel shortage was a reduction in the area under irrigation, and in 1998 this area was still down by 22,000ha compared to a decade previously (FAO, 2004). Overall, oil and agrochemical imports were increasing slightly over the decade, but quantities were fluctuating and still vastly lower than in the 1980s.

Substitution with organic production techniques and system innovations

On the one hand, all available petrol, fertilizer and pesticide inputs had to be distributed for maximum benefit. This meant prioritizing the most important crops for the receipt of inputs, such as sugarcane and potato, and also banana, rice, tobacco, citrus and certain horticultural crops (FAO, 2003). Similarly, irrigation was concentrated on a very limited number of farming units (Paneque Brizuelas, 1997). On the other hand, the shortfall of petroleum-based imports had to be substituted by domestically produced alternatives. This meant introducing and developing the use of biological pest and disease controls, organic soil fertility strategies and draught power, and accompanying training.

A national programme for biological pest control had already been established in the late 1980s. Within this, two main approaches were used to control insect pests: the release of beneficial insects (entomophagens), which acted as parasites on the eggs of pest species; and the use of natural bacteria and pathogens of certain pest species (entomopathogens). Over 220 small laboratories and production centres, or CREES, were constructed nationwide, for the production and distribution of these biological pest and disease controls. Using these two strategies, scientists developed techniques to combat the main pests of major crops, including rice, sweetpotato, sugarcane, cabbage, tobacco, coffee and citrus (Sinclair and Thompson, 2001). Figures on the actual extent of usage of biological pest controls were scarce, although production figures of the major controls – *Bacillus thuringiensis* and *Trichoderma spp* – showed a significant increase between 1994 and 1997 (Díaz, 1995; CNSV, 2000). The 1990s also saw large-scale efforts to recycle nutrients and to use available sources of organic materials (FAO, 2003). The main organic soil amendments being promoted were manure, compost, worm compost, sugarcane wastes, green manures and biofertilizers. However, and similar to pest controls, figures on actual usage were scarce.

Rural innovation programmes were developed and expanded to 22 programmes by the end of the decade. These were led by Raul Castro. One of the first was the use of draft animals, established in 1992 through the Ministries of Agriculture and Sugar. It became obligatory to deliver all fit bulls to cooperatives and state farms. This provided 100,000 bulls, and with these a breeding programme built up numbers to 376,000, from which almost 30,000 oxen were being trained annually. Training in the handling of oxen teams was rolled out. Early challenges to this programme included a lack of farm infrastructure to support animal traction, lack of relevant knowledge by the agricultural workforce, lack of pasture and on-farm fodder supplies (compounded by problems of transportation of feeds), and insufficient veterinary services (Ríos, 2002). Between 1990 and 1997, total oxen numbers doubled, although these were largely being used on small private farms. At 1998, the private CCS and individual farmers, working 15 per cent of the land, were using 78 per cent of draft animals (Ríos and Aguerreberre, 1998).

Another innovation programme promoted farm diversification: all farms that were previously dedicated to a few commodity crops or to livestock were now obliged to become self-sufficient or diversify for the market. Rosset (1998) notes that, whilst the use of organic inputs as a replacement for agrochemicals was important for maintaining production levels, two other factors had contributed to increasing the complexity and diversity of agricultural systems: intercropping; and the integration of crops with animal production in rotational systems.

Adapting agricultural education, training and support

The revolutionary state had for many years placed importance on education for the agrarian sector. In the 1990s, a National Commission for Extension Education was also formed, and management staff of most cooperatives had, by the end of the decade, received some form of management and/or agronomic training. *Campesino* farmers continued to receive training at the Niceto Pérez National Training Centre of the National Association of Small Farmers (ANAP), in Havana Province, and this centre – along with a range of Cuban NGOs including the Council of Churches of Cuba (DECAP), the Cuban Association of Small Livestock Producers (ACPA), the Cuban Association for Organic Agriculture (ACAO) and the Foundation for Nature and Humanity (FANJ) – developed training courses in agro-ecology. ANAP was also promoting agro-ecology through its farmer-to-farmer training approach, which was introduced into Cuba in 1998. In 1994, the National Agricultural University (UNAH) opened the Centre for Studies in Sustainable Agriculture (CEAS), which offered part- and full-time and long-distance diploma and MSc courses in agro-ecology. ACAO assisted in course development, and the long-distance course was aimed not only at teachers, scientists and agronomists, but also at farmers (Rosset and Moore, 1997). In terms of specific education on organic agriculture, ACAO worked to diffuse information through its regular publication, *Agricultura Orgánica*, and also through the operation of mobile libraries that travelled around the country.

In the early part of the 1990s, an interest developed in the potential of traditional knowledge as a way of helping Cuba out of its Special Period. Rosset (1996) describes how a new programme by MINAG encouraged farmers to share their knowledge with researchers and officials at mobile workshops throughout the country. Exchange of traditional knowledge between farmers was also encouraged by ANAP. However, this interest had reduced by the end of the decade.

Technical assistance and support from the state was historically paternalistic, more than simply providing information and services. Support came mainly through the municipal offices of MINAG, the Agricultural Enterprises, which for historical reasons tended to focus on the state farms and UBPCs. Private farmers were catered for by the Campesino Cooperative Sector (SCC), which, for example, distributed work clothes and tools provided by MINAG, and which liaised between the farmer and Sanidad Vegetal at the appearance of a pest or disease. ANAP, with offices in every municipality, also guided the *campesino* cooperatives, with one ANAP representative responsible for 8–12 cooperatives.

State reforms in farm organization

Because of the historical rural-to-urban flow of migration, labour shortages had been a longstanding problem on farms. Changes in farm organization were considered necessary to address this, as well as to address the low productivity and inefficiencies of centralized planning in the state farm sector. These changes involved restructuring farm size and organization, improving access to land, increasing of rural wages, and creating new agricultural markets.

The dismantling of large-scale farms

From the start of the Special Period there ensued a dramatic reorganization and redistribution of state land in favour of the growth of cooperatives, as experience showed cooperatives to be more productive, especially in a low-input situation (Deere et al, 1994; Deere, 1996, 1997). Even Castro admitted, 'The state has not had success in large farm business' (Granma, 2003). Rosset (1996) expands on this theme: 'Once considered a national embarrassment, a remnant of a backward past, peasant farmers are emerging from this crisis with their image revamped by their agile response.' One completely new type of production system was instated, and other state farms and CCS cooperatives were structurally reorganized or strengthened. Whereas, in 1992, 75 per cent of cultivable land was under state control, by 1997 this figure had dropped to only 34 per cent (Fernández, 1998). This change in land management is shown in Figure 7.1.

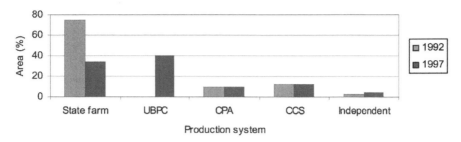

Figure 7.1 *Area occupied by each type of production system, 1992 compared with 1997 (% of total farm land)*

Source: Fernández, 1998

Changes in state farm structures

The majority of state farms were transformed in one of two ways. Basic Units of Cooperative Production, known as UBPCs (Unidades Basicas de Producción Cooperativa), were formed in 1993 out of existing state farms. Land was handed over in rent-free perpetuity (*en usufruct*) to the state farm workers, the aim being to stimulate production in the style of the long-existing CPAs, without privatizing to the extent of the CCSs. In this transfer, the workers became associates and had to purchase the farm's capital infrastructure. Members had long-term credit arrangements, they owned the produce and received an equal share of profits, and they could elect their own management teams (Deere, 1997). Decision-making on production was decentralized, yet centralized planning was retained for overall resource management and food distribution (Sinclair and Thompson, 2001). By 1995 there were 1353 UBPCs, covering 37 per cent of Cuba's agricultural area (Everleny and Marquetti, 1995) and incorporating 400,000 ex-state agricultural workers (Lage, 1995). Agricultural labourers accounted for 95 per cent of the members. Most state sugarcane farms were converted in the hope of raising sugar yields. In terms of performance, these UBPC farms were very slow to take off; towards the end of the 1990s they were still considered to be economically inefficient and had not yet achieved the results hoped for by the state. Many of these disappointing early results could be attributed to the continued maintenance of old forms of power and administrative structures (EIU, 1997).[1]

The other type of state farm transformation was to New Type State Farms. Known as GENTs (Granjas Estatales de Nuevos Tipos), these comprised farms that were considered unsuitable at that time for conversion to UBPCs. As one farmer put it, 'We want to convert to a UBPC but we first need to show that we are profitable. The process is slow.' These farms continued to produce on a large-scale and industrialized basis, and were operated by the state. However, several changes were made in their organization: they were granted more decision-making power, more economic responsibility and better representation of the personal interests of their workers. As with state farms, 50 per cent of profits were distributed amongst the workers, including management staff, 25 per cent went into investment and capital growth, social development and contingency reserves, and 25 per cent to the state budget. On top of the fixed salary, workers could obtain bonuses by producing a surplus for sale through the new farmers' markets. The average monthly income was $13 (254 Cuban pesos) (AGROINFOR, 1998).

Changes in non-state farming structures

Within the *campesino* sector, the main change was in 1997 with the development of 'strengthened' Credit and Service Cooperatives (CCS-fortalecida). This move to build the capacity of certain CCSs was based on the recognition that they had outperformed both CPAs and state farms, with far less state investment (Sinclair and Thompson, 2001). 'Strengthening' consisted of fusing two normal CCSs to cooperate over services. The new cooperative could draw on a better range of professional services, and could also make a production plan with Acopio as one entity rather than with each individual farmer member. By 1999, approximately one-third of CCS cooperatives had been strengthened in this way, although some CCS members were concerned that these changes were pushing them closer to the CPA model (Miedema and Trinks, 1998; Sinclair and Thompson, 2001).

Overall increase in diversity of farming types

By 1997 there were 1500 UBPCs occupying 21 per cent of agricultural land, 1150 CPAs occupying 9.4 per cent, and 2700 CCSs occupying 12 per cent. In addition, 71,000 individual workers were holding 103,334ha of land in perpetuity (3 per cent of agricultural land), mainly for coffee, cocoa and self-provisioning. These non-state farms became responsible for 90 per cent of sugarcane land and 42 per cent of non-sugar cropping land. Although they comprised the UBPCs, CPAs, CCSs and individual producers, only the CPAs, CCSs and individuals were classified as *campesinos* (Wroe, 1996; ONE, 1997; MINAG, 1998).

In addition to the remaining state farms, there were 62 New Type State Farms (GENTS), 75 farms operated by the Youth Work Service (Ejercicio Juvenil de Trabajo, EJT), and 19 farms operated by MININT and MINFAR.

State reforms for incentivizing production

'Linking man to the land' (vinculación del hombre al area)

In 1990, a new incentivization system was introduced on farms, a 'Technical, Organizational and Payment System According to Final Production Results'

(*Sistema Tecnólogico, Organizativo, y de Pago por los Resultado Finales de la Producción*). This system was introduced onto state farms in Havana Province, and then extended – albeit unevenly and without obligation – to all MINAG farms around the country, including UBPCs and CPAs. Under this system, an individual worker or group within the farm was responsible for a specific area of land and its production – typically four people to 13.4ha – and this individual or group received a share of any surplus to the production plan. Conversely, a share might be deducted from the wage packet if production costs were too high and targets not met (Enríquez, 2000).

This system was in contrast to the previous labour management structure, whereby individuals or gangs of workers would be rotated around the farm unit and paid on the basis of days worked. The new system was designed to increase labour productivity and efficiency, in order to redress a situation where, according to Rosset (1996), 'The bond had been cut between the farm worker and the land. In crisis, state farms' unwieldy management units could not adapt to life without high inputs of technology.' The effects of this new system on productivity were difficult to assess early on, as its introduction coincided with a decline in inputs and implementation was found to be a challenge for short cycle crops or where voluntary workers were involved (Deere et al, 1994). As one Cuban economist (Fernández, 1998) explained about this process, 'The first thing we want to do is to turn workers into farmers, to make them feel like owners and then to produce as the *campesinos* do.'

Provision of land in perpetuity

In September 1993, MINAG decreed that small farmers and CPAs were allowed to receive more land in perpetuity if they used it for the production of coffee or tobacco. This specifically applied to old state land that had fallen fallow during the early years of the Special Period. Distribution was based on contract and quota, and as long as this was met then the land could also be used for other commercial activities. Unproductive land could be claimed back by the state. In addition, land not exceeding 0.24ha was also made available in perpetuity to individuals who were unable to work full-time in agriculture, and to pensioners, in order to increase family self-provisioning. 'In perpetuity' meant rent-free leasing, with the state holding the property rights (Rosset, 2002). This move demonstrated flexibility on the part of the state, which had to discard some long-held principles of opposition to land held in perpetuity (Deere et al, 1994).

Change in farm-gate prices and markets

In the 1990s, changes were made to the system of planned production for the ration, operated by Acopio. Traditionally, the state collected an agreed quantity of produce from each farm or cooperative, at an agreed, low price. Changes were introduced whereby basic prices for produce were still determined at the time of planning, but any surplus production by the farmer now received a slightly higher price from the state (sugar was an exception to this). The farmer also now had the choice to sell this surplus through other channels. For export crops such as sugarcane and tobacco, part of the purchase price was paid in US dollars. Individual farmers received this in the form of cash, though in the case of cooperative members the cash was kept in a state account that members could access to purchase dollar goods such as tools, inputs and clothing.

Marketing arrangements were diversified in several ways, as discussed in Chapter 6, but two were made specifically to incentive farmers. One was the short-lived enabling of direct sales or 'tiro directo', through an agreement between a farm and local sales centre. The other change was, in 1994, the opening of farmers' markets throughout the country.

Farmers' coping strategies

It might be assumed that the centralized planning system in Cuba would lead to homogeneity within the farming sector. This was not the case; there were significant differences at all levels: between crop management approaches, between farms and between regions. Because of their varying locations, histories, management structures, resource, knowledge levels and so on, each farm was unique, and attempts at generalization lose much of the detail that actually provides the means to understand the situation.

The reality of agrochemical application

The reality of agrochemical use was more complex than might be assumed. Although import levels did drop at the beginning of the 1990s to a low of 25–33 per cent of previous, farmers were buffered by reserves, so that the most difficult period in terms of agrochemical availability occurred in the mid-1990s when these reserves had run dry. One UBPC farmer explained: 'In 1993 to 1994 we still had inputs and reserves including spare parts. So we were profitable for two years. It was in 1995 that yields fell – up to 1997 were the most difficult years.' The majority of farmers did experience a decrease in agrochemical usage, although these experiences were far from uniform. Some farmers, and especially *campesinos*, had never been large users of agrochemicals and could maintain similar levels as previously, or even increase them as incomes rose sufficiently to be able to purchase more on the black market. Two CCS farmers explained: 'I estimate that we are using the same amounts as before the crisis, but the difference is that previously it was much easier to obtain from the state Enterprise,' and 'In fact we do use the same levels as in the 1980s, but in any case it was never our custom to use very much at all because we are traditional producers and so we only plant on a very small scale.' The black market served as a longer-term buffer to eke out reserves and spread them via the wealthier *campesino* farmers. One UBPC farmer explained: 'We borrowed inputs from the *campesinos* who had more reserves.' Other farmers who had previously had little access to agrochemicals found themselves incorporated into the modernization programmes of the 1990s that actually made these inputs available for the first time. For one UBPC farmer: 'Now yields have increased because there's more availability of petrol, chemicals and pesticides, because of agreements with other countries and because of national research.' For the rest, inputs started to become more available towards the end of the decade. Overall, farmers exhibited three types of response to the lack of agrochemicals and their approach to its use, as shown in Box 7.1. These attitudes were, for many, the same as held prior to the Special Period.

Box 7.1 *Farmer types in relation to attitude towards agrochemical usage*

Farmer attitude	Approach towards agrochemical input use
Frugal	Conservative over input use, is responsive to crop and environment
Never enough	Always accepts more agrochemicals if available, though would also use frugally and store reserves, driven by feeling of insecurity
Unnecessary	Does not believe in using agrochemicals unless absolutely unavoidable, usually trained in or has encountered organic concepts

Source: Wright, 2005

A patchwork effect of practices had emerged across the country. Agrochemical input distribution and access was officially controlled by the state. Farmers required permission in order to purchase agrochemicals from official sources. Each municipality had different priority crops, and certain provinces, such as Havana, were more highly prioritized than others. High-performing farmers, including those with good natural resources or irrigation equipment, and increasingly the *campesino* farmers, received preferential treatment. In contrast, only 18 per cent of the total area sown with non-prioritized crops received chemical fertilizer in 1998 (FAO, 2003). This created a patchwork effect of agrochemical application, in both dosage and area covered, at farm, municipal and provisional levels, and, accordingly, a patchwork of alternative measures also.

Owing to the fluctuations in national input availability, in practice even prioritized crops and areas could fall short of resources. Input prices were also variable. Farmers had different experiences of input price rises, estimating variously as between 60 and 300 per cent since the 1980s. For example, chemical fertilizer that previously had cost $3.6/t was currently selling at $11–$25/t on the black market. Agrochemicals were on average thinly spread. Dosages of nitrogen-based fertilizer were quoted as low as 0.11t/ha, and pesticides from 0.06l/ha. Usage was often as a precision application, or as a curative rather than preventative. Farmers had learned to use agrochemicals more sparingly and efficiently, and to take into account agroclimatic influences such as rainfall. Nevertheless, 75 per cent of the cooperative farms visited were regularly using agrochemical fertilizers on at least some of their crops, and 65 per cent were using pesticides, herbicides and fungicides. There was also a widespread desire to use more agrochemicals, especially fertilizers, when they became more available – expressed by 83 per cent of farmers interviewed across the country.

More important than agrochemicals: fuel shortages and adequate germplasm

The system of prioritization was also applied to fuel distribution: different regions, crops and farmers had different levels of priority. This affected both machinery and irrigation use. For farmers, the lack of fuel for irrigation was more of a limitation to production than of agrochemicals, and the most common complaint was of the lack of petrol (*falta petroleo!*). For some, such as the following CCS farmer, the lack of petrol was a limitation that they could not see their way around: 'There's nothing else I can do really, if there's no petrol, because in the winter there's no rain and then no irrigation either, and there's no point in applying inputs if the soil is not humid. I've heard of other farmers who produce with no petrol – in Peru – but they must have different soils.' The crucial importance of fuel for irrigation was substantiated by another farmer: 'We do lack fertilizer but we can't use it if there's no irrigation.' From the farmers' perspective, fuel supplies remained low over the decade, generally meeting only 20–60 per cent of felt needs. Irrigation was also rationed according to crop: sugar took 39 per cent, roots, tubers and grains 21 per cent, and rice 16 per cent. Only one farm encountered was experimenting with solar-powered micro-jet irrigation. Although petroleum prices had doubled since the 1980s (in Havana, for example, from 9c/litre to 20–25c/litre), the cost of fuel was not in itself felt to be a constraint, and 70 per cent of farmers interviewed would use more farm machinery if only they had better access to fuel. These shortages meant that farmers learned to use whatever fuel they acquired far more sparingly and effectively than in the 1980s, and the allocation of available fuel to various activities was the major topic of the cooperatives' weekly planning meetings. Figure 7.2 depicts farmers' responses on the perceived main input constraints to production.

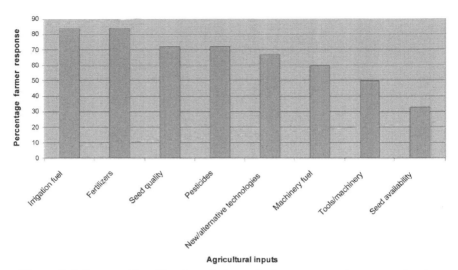

Figure 7.2 *Farmer identification of the main input constraints to production, 1999–2000 (% response from sample)*

Source: Wright, 2005

Farmers rated access to adequate seed varieties and seed quality as being as important as, if not more than, access to pesticide inputs. Whilst farmers had reasonable knowledge about seed management, they held very little autonomy over seed selection and production, and on-farm seed saving was limited by the lack of adequate local storage. The state seed sector, which was historically in control of selecting, reproducing, storing and disseminating seeds over the country, remained centralized during this period. Yet towards the end of the decade its ability to adequately fulfil its functions was very weak, and this resulted in farmers having to use inappropriate varieties and reduced levels of diversity. The country was dependent on seed production in Havana or from abroad for half its seed sources, and there was a lack of resources to develop new hybrids domestically. This situation was spawning a black market for seeds, and intermediaries were travelling throughout the country, buying and selling.

The extent of use of organic techniques and resulting land use patterns

Farmers were forced to look to on-farm resources and the natural resource base in order to maintain or increase production. Because of this, many alternative coping strategies were consistent with organic principles, albeit unintentional; they were something that farmers had to do by default. These strategies could be broken down into those relating to soil fertility, to pest and disease control, to water usage, to traction, and to varietal choice and seed management. Appendix A provides data on the degree of use of organic strategies in these categories.

Soil fertility coping strategies

Soils were generally in poor condition. To improve soil fertility, and along with chemical fertilizers, almost all farmers were, on some part of their farm, rotating crops, incorporating crop residues into the soil, and using oxen. A large number were also intercropping, practising minimum tillage and applying manure (at rates of between 0.1 and 22t/ha). A minority of farmers were using compost, legume crops and worm humus, some of which was produced on-farm. Terracing was practised to a small extent in hilly regions. Rotations were planned to take advantage of chemical fertilizer residues remaining after harvesting of a prioritized crop.

Off-farm manure came from state livestock farms, sugarcane waste (*cachaza*) from the Agro-Industrial Centres (CAIs), and green manures and biofertilizers from research institutes. In some cases, obtaining free manure had become more difficult, as the specialized livestock farms had themselves diversified into crop production for which they used their own manure, or had started to charge for this product as a result of more efficient management. Prices and access to manure were also dependent on transport availability.

Pest and disease control coping strategies

As well as chemical controls, almost all farmers were using or had used a range of other pest and disease control methods. These methods included selecting resistant crop varieties where possible, using natural weather patterns to disrupt pest lifestyles, applying the natural pest control *Bacillus thuringiensis* (Bt)[2] and rotating crops. Biological control products came from the CREEs. A large number were also applying other biological pest controls, and using a mixture of simple, good management practices. These included the manual removal of pests and eggs, mulching, removing infected leaves and maintaining clean crop rows, good soil preparation, separating susceptible varieties, irrigation and the use of clean seeds. A farmer on one CCS located in the mountainous eastern region was intercropping banana with coffee, and explained: 'When there's more sun, there are more pests, so we intercrop the banana with coffee to provide shade, and it makes a difference.' Another had integrated a series of practices: 'The secret of high yields is to take care of the crop and the seed. I've just used more care and attention now. I have the highest cassava yields in the country – 20t/ha, when the provincial average is 10. I apply *Trichograma* – it's easy to use as the plant health technician decides when to apply it – in the mornings, and one packet is sufficient. Hoeing is required now but it doesn't need much labour because we just changed strategy to hoe when the weeds are smaller – this is more efficient.' Biological control inputs were also used in a patchwork effect; most commonly by UBPC cooperatives. In fact, CCS cooperatives appeared to use these inputs to a minimum, as was the case for all farmers in certain regions of the country. Overall quantities, frequencies and extent of use of biological control inputs were difficult to quantify, although secondary data supported the evidence that usage of biological pest controls was in practice actually quite low.[3]

Water usage coping strategies

Over half of the farmers were irrigating at least part of their farms, albeit at levels termed 'survival irrigation', with equipment mostly driven by petrol-driven turbines or by gravity. Some *campesino* farmers had been able to invest their new wealth in electric turbines. All farmers were synchronizing crop sowing with the arrival of the rainy seasons, combining awareness of planting time, rainfall variation and soil type, and all had adapted their irrigation schedules to make them less wasteful. Maize for example was cropped twice a year, and earlier planting was now dependent on early rains. 'Previously we could plant anywhere, but now this is not so,' explained a CCS farmer. Almost all were using drought-resistant crop varieties where available.

Many farmers were attempting to adapt their irrigation systems. In fact, this was the most common technological adaptation that farmers were making, not least because it was a crucial prerequisite for applying chemical fertilizers as well as some biological controls. One described the change: 'At the crisis I had to invent. For my rice, a water expert showed me how to adapt my tank for gravity irrigation. Family in the Canary Islands passed on the technique of creating a cascade and reservoir to irrigate the land below – its an old technique from the beginning of the century and now in the 1990s it's being used more again.' Another strategy was to maintain clean irrigation ditches: 'It's important to keep channels free from weeds. I always did this but now I've improved the efficiency – by hand is most effective, but it can also be done by tractor or oxen.'

Traction coping strategies

Levels of farm mechanization varied, from a CCS farmer with one pair of oxen for all farm activities to a UBPC with 50 oxen and three tractors. Many activities were still carried out by hand, and hand labour had increased, in particular for cultivation and harvesting. All farmers were using oxen on some part of the farm, for supplementing rather than replacing machinery, and had modified farm equipment for this purpose. One example was the use of a Cuban innovation, the 'multi-plough', which had been designed so as not to invert the topsoil. The use of oxen was seen as a complement to mechanical approaches, but not a replacement for them. Tending oxen was seen as hard work – they also required guarding from theft, and there was little development of support for their maintenance, such as blacksmiths, fodder sources or specialized veterinary services.

For some farmers, the lack of fuel to drive machinery caused land to be insufficiently prepared. Others, however, noted that the use of oxen had led them to pay more care and attention to soil management. Labour savings were made by leaving crop residues in the field, practising minimum tillage and sharing the scarce machinery available. In some cases, fencing and wild areas were being taken up in order to facilitate tractor manoeuvring. A widespread concern amongst farmers was that the deterioration of old farm machinery, and lack of petrol, had adversely affected agricultural productivity.

Varietal choice and seed management coping strategies

Choice of seeds and planting materials were affected by the market, as well as by economic circumstance and agronomic conditions. Although most farmers obtained at least some seed from the state Seed Enterprise, at least 20 per cent of this did not germinate. CPA and CCS cooperatives also obtained seed from neighbours and through self-saving, although seed saving practices had in some areas decreased, owing to the incidence of theft if seed crops were left out to dry, and to challenges associated with seed storage. One CCS farmer commented on changes in seed sourcing: 'We used to exchange seed with our neighbours from other cooperatives, but with the introduction of the *Plan de Semillas* [state Seed Plan/Enterprise], this wasn't necessary any more.' Some cooperatives attempted to store small quantities of seed as a security measure, while others stored the seed they used for self-provisioning. Farmers were unable to travel to prospect for seed in the same way that they had been used to. Drawing from traditional knowledge, *campesino* farmers knew how to select for desired characteristics within certain crops. Modern crop varieties were valued by farmers for their yield potential but, beyond that, traditional varieties were believed to be superior in terms of flavour, cooking quality, pest resistance, drought resistance, hardiness in poor soils and wind resistance. Resistance to drought was particularly valued as this meant fuel savings could be made through reduced irrigation schedules. Approximately half of the farmers felt that modern varieties and hybrids required higher levels of inputs, while the rest saw no difference. With the Special Period, the state Seed Enterprise had been forced to give up its own seed multiplication areas and contract out farmers instead. This change had led, according to one CPA farmer, to the loss of the local varieties that these farmers had previously grown.

Changing land use patterns

Because of the lack of inputs, land use patterns were changing. The lack of irrigation limited the use of certain land for crops, and with the pressure to intensify, so more fallow land was being brought into production. The increase in rotations was having a stronger influence on cropping dates and locations than input availability or soil type. Crop types were changing to those that were less input-hungry or that were more bulky – such as roots and tubers – to more easily fulfil the tonnage-based state production plans. One CCS farmer explained: 'In the 1980s we planted less maize and more taro, which required more inputs, as did carrot and cabbage. Since then we've had to plant more roots and tubers as these require fewer inputs, although they have lower yields.' The state was prioritizing certain crops over others, and urban agriculture was popularizing specific items. The risk of theft also influenced what was planted particularly around farm and field borders. Legume production was increasing as a national priority for self-sufficiency rather than for improving soil fertility, and livestock numbers had decreased due to the lack of feed. Finally, as farm sizes decreased (in the case of one farm, for example, from one 1000ha unit in 1993 to four smaller UBPCs by 1998), so regional diversity was increasing.

Strategic differences between CCS and UBPC cooperatives

Strategic differences between cooperative types were apparent. Because CCS cooperatives could generally access agrochemicals on the black market, so these farmers were less 'organic' than might be supposed. They used traditional organic inputs such as manure, but did not feel the need to use the more recently introduced biological pest controls and bio-fertilizers. One CCS farmer explained: 'We CCSs can afford these black market prices, but the CPA and UBPC cooperatives cannot, and this is the reason they turn to alternatives.' UBPCs tended to use more of these organic inputs, as well as more intercropping and minimum tillage practices. They were more influenced by state programmes such as for the promotion of biological controls, but found the high costs of manure to be a constraint. UBPCs were serviced with seed supplies from the state Seed Enterprise, whereas CCS farmers used more traditional varieties and self-saved seed. Again because of differences in income, CCS farmers were able to purchase modern irrigation equipment, although this did not prevent UBPCs from irrigating larger land areas. When it came to traction, CCS farmers used more oxen and also had a higher labour density than that of UBPCs. Taking Holguin Province as an example, the CCSs irrigated 19 per cent of cultivable land and had an average work density of 28 persons/ha, compared to 31 per cent of irrigated land for the UBPCs and a work density of three persons/ha.

Overall, if one were to summarize the characteristics of the farmers most satisfied with their current situation – those who complained the least – they all owned electric irrigation pumps, their soil fertility strategy was based around the application of organic matter, they owned or had easy access to a tractor, and they applied only small quantities of agrochemicals.

Farmers' preferred future production strategies

The majority of farmers (83 per cent) expressed the desire to use more agrochemical fertilizers, and two-thirds would use more pesticides, as and when they could get access to them. The main reason was to increase yields but also because of ease and rapidity of use. Sixteen different pesticides and herbicides were named as being desirable. One farmer explained: 'My chemicals work fine. I don't need to try anything else. When I want to buy any chemicals, I simply get authorization from the CCS technician.' For some crops and diseases there was felt to be no alternative to chemical control. 'When there's a plague you have to use them' and 'I have to use herbicide because the stony soil means that I can't work it' were two responses. Others reported that 'This technology is necessary to develop the country,' and 'Every day there appears a new, stronger and unknown pest.' There was also uncertainly over alternatives; some farmers were concerned that manure and mulch might contain weed seeds or hold insufficient crop nutrients. Others felt that biological pest controls were not fail-safe and therefore carried too high a risk. There was a also a fear that there would not be enough organic inputs to go round. However, very few farmers would prefer to use only agrochemicals, and the majority (83 per cent) would also use more organic inputs if they could get access to them, largely because of awareness over the negative impacts of agrochemicals, and many farmers were organizing to produce their own manure and worm compost supplies. Biological pest controls were a popular option. 'Biological control is better – it's more economical, more available, less toxic and has been effective,' explained one farmer.

For those farmers who had no intention of using more agrochemicals, some were satisfied with the low levels they already had, whilst others emphasized knowledge as the key to farming without chemicals. One explained: 'Most farmers apply too much fertilizer because they don't know their soils and have not had a soil analysis undertaken.' Whilst another said: 'To improve yields, what we need is a dairy herd, compost and more traditional knowledge.' Some farmers would prefer to use only organic and felt that this was more appropriate, as two farmers expressed: 'the use of alternatives means less costs, less pollution and less external dependency', and 'this will result in resistant plants with low input needs'. Overall, however, an integrated approach was the preferred strategy, for various reasons including the delay of onset of pest attack, improved germination rates and prevailing poor soil quality.

In terms of irrigation and traction, all farmers would make more widespread use if they could get access, although they would modify equipment to be more efficient. Electric turbines were the most popular choice of desired technology, as those farmers already with this equipment had shown themselves to be relatively unaffected by the decrease in petrol availability. Very few farmers were aware of or interested in alternative water management systems. More farm equipment was also desired, with very few farmers preferring to use oxen. One exception to this explained his preference: 'Oxen are better for the soil, they do not require much more time, they are more precise and attain a higher quality.'

Farmer knowledge levels and technical assistance

Along with input availability, another factor influencing coping strategies was the level of on-farm knowledge. CPA and CCS cooperatives contained a substantial number of farmer-members trained in different technical disciplines, as well as members from other professional backgrounds. Amongst members there was also a high level of traditional knowledge that had survived the industrialized period. As one older farmer explained, 'I have always been a *campesina*. I left school at seventh grade to work on the land with my grandmother, and now I live on my son's land.' This knowledge was highly under-exploited on the CPA cooperatives in particular, where older farmers felt their knowledge to be outdated. This feeling was exacerbated by the introduction of 'new' organic and biological inputs such as Bt and worm humus. For example, one elderly CPA member had worked in agriculture for 50 years and used to run his own farm before pooling it into the cooperative, yet did not take a major decision-making role in planning and management. He explained: 'I am only a *campesino* so I am still learning about the new methods of chemical control and even newer ones of biological control, I am still trying to align the new approaches with what I already know.' Box 7.2 provides an example of the latent traditional knowledge still held on-farm, whereby one older CPA member was asked to describe the traditional production of maize in Cuba.

The CCS cooperatives did not pool their knowledge and expertise to the same extent, and CCS farmers noted the usefulness of traditional knowledge during the Special Period. One explained: 'I refer back to traditional knowledge more than before – for crop varieties, practices such as rotation, and so on, though some things are just not possible now because of the [state production] plan.' Compared to UBPC cooperatives, CCS cooperative members were on average nine years older, held twice as much agricultural experience, and spent four times as long on the same cooperative. With a higher turnover of migrant and uneducated workers, the UBPC cooperatives held less traditional and local knowledge, though there were attempts to address this through training. As the head of one UBPC explained about his cooperative's expansion, 'The more recent members have come here through the granting of land in perpetuity – they have very little agricultural knowledge, they are just workers [*obreros*].' Unskilled migrant labour tended to gravitate towards Havana, and this showed amongst UBPCs that were located further from the capital whose members were more mature, local and experienced, and more committed to longer-term sustainable strategies.

Notwithstanding the training programmes available, knowledge levels on alternative, organic approaches were still relatively low at the end of the decade. Farmers most commonly received training on bio-pesticide use, and a smaller number on bio-fertilizers. Other topics approached from an alternative

---

Box 7.2 *Traditional methods for maize production in Cuba, as described by an elderly CPA cooperative member*

In the past, maize was selected for the fullest cobs that had the tightest wrapping of husk, with no hole at the end for insects to penetrate. On a night when the moon was full but about to wane, the seed was harvested and bagged in jute sacks. Weevils would not eat the grain when the moon was waning (*menguante*). The jute sacks had been used for sugar and they kept the cobs better than nylon. Fifty cobs were put into each sack, and these then strung up in a dry, well-ventilated place. There they would keep fresh for 5–6 months. Only the seed from the middle of each cob would be sown, because at the cob-ends the seed was considered too weak. Sowing periods were planned to take advantage of the rains, and germination rates were almost 100 per cent. The seeds would be sown in an open row with about 18 inches spacing between plants, and 2–3 grains per hole. Then the row would be closed by foot. When the maize was tall, the plot would be weeded and the rows hilled-up. All cultivation was undertaken with oxen.

The farms used to be mixed, and farmers would rotate crops with grazing pasture. The soil was fertile and did not need fertilizing directly, but prior to the maize they would grow sweet-potato or cassava and leave the residues in the field. This was called 'vegetal al terrano'. After ploughing, the soil would be left to rest for 40 days before sowing. During this time, the weed residues would break down and fertilize the soil, and because of the heat it was said that the broken soil had 'fever'. If crops were sown before the end of this period they would turn yellow. The maize in those days was *criollo* – locally interbred and adapted. The cobs were yellow and bottle-shaped. Maize harvested at three months was tender (*tierna*), while that harvested at four months would be dried, ground and made into a dish of steamed maize (*tamales*).

Source: Wright, 2005

---

perspective were negligible, and there was little awareness on the underlying philosophy of the organic approach, as opposed to a simple toolkit of organic technologies. The majority of farmers who did hold ecological knowledge had learned it from family or through experience.

Farmers received relatively high levels of technical assistance, most commonly for pest and disease control, and then for seed and soil management. Fewer farmers received advice on irrigation or processing and marketing. Table 7.1 shows the most common thematic areas of technical assistance received by farmers.

UBPC members received significantly more advice on these topics. For all farmers, technical assistance was free. The degree of appropriateness of advice was variable. One CCS farmer explained: 'The technical expert from the Seed Enterprise gives us advice on everything to do with plant health, with a focus on the use of chemicals.' Soil analyses frequently had to be paid for, but over half the farmers had undertaken these. This was a marked change from the

Table 7.1 *Percentage of farmers receiving technical assistance by thematic area*

Theme	% farmers receiving technical assistance
Pest and disease control	80
Seeds	74
Soil cultivation/mechanization	68
Soil/crop fertility	65
Irrigation	47
Processing/marketing	42

Source: Wright, 2005

previous decade, as explained by one farmer: 'Before the Revolution, *el quimico* [the chemical expert] came every year to analyse the soil and provide inputs. After the Revolution, the soil wasn't analysed, and in the 1970s and 80s we just added more and more inputs.'

Each cooperative had at least one full-time technical support staff. Previously, advice had come from the organizations based at provincial level, but recently, farmers noted, more locally based institutes were active in providing support. For CCS farmers, a major change had occurred in 1996, when the local Enterprises of MINAG had taken over from the Campesino Cooperative Sector (SSC) and ANAP in providing support and input distribution. As a result of this change, inputs and resources had almost ceased to flow to *campesinos* in certain regions during the period 1997–99, and this, coupled with the drying-up of their own reserves, meant that the second half of the decade for them had been arguably worse than the first half. One CCS administrator explained: 'I've worked in the private sector since 1987, and I've seen the changes over the years. I think the private sector suffered a lot during the 1990s, but the biggest impact was in 1997 when it had to start obtaining assistance and resources from the Enterprises, because previously it had its own separate source of assistance.' The specialized nature of the Enterprises meant that they were not skilled for the task, and in 1999 the SSC was reinstated and farmers immediately noted the positive results. Towards the end of the decade, and with the realization that the UBPCs required more intensive training in order to convert workers into responsible farmers, so they began receiving more support from the Agricultural Enterprises. This positive increase in support was described by one UBPC farmer: 'From 1995 to 1999 this was a state military farm which received lots of inputs but no technical knowledge. Now everything has changed. We have one-day courses and strategy plans organized by the Enterprise.'

Box 7.3 *Coping strategies for farm management during the Special Period*

Rationalization

Farmers learned to be more effective and productive through more rational usage, not only of inputs but also of labour, and more 'care taking', with fewer losses.

Increased responsibility

Particularly for UBPCs that were converted from state-managed farms into worker cooperatives. As one UBPC farmer noted, 'Previously everything was given to us, but now we have to look for it and invent it.'

Improving incentives

The cooperatives themselves worked to provide better conditions and incentives for their members, especially on the UBPCs.

Self-sufficiency

For CCS contract workers and UBPC members, self-provisioning was introduced and increasing as an additional incentive and survival measure. One farmer noted: 'I intercropped banana with cassava to ensure a harvest of some staple food in case of drought.'

Intensification

This increased efficiency, in terms of more harvests per year, more short-season crops, more intercropping and synchronized rotations.

On-farm diversification and complexity building

Farms were expanding their range of produce and developing on-farm connections so that by-products from one activity could be used for another, such as mixed crop–livestock farming. Especially UBPCs were venturing into 'aquatic livestock', apiculture, lombriculture, as well as production for the tourist market. As one farmer explained, 'We've survived by two strategies: better market prices, and diversification. We've had an *organoponico* since 1993. We have more production with fewer costs – fewer inputs, more manure and more efficient use of irrigation.'

Targeting niche markets

By focusing on local human skills and on the biophysical potential of the land, some farmers were receiving state incentives for niche production such as for early or late-season crops. In the case of one CCS farmer, 'I plant maize, tomato and bean out of season by taking care and preparing the land well – so they are all high value,' and another: 'Our survival strategies are linking man to the land, looking for short-term crops of high yield and high value such as flowers, and the use of bio-inputs to control pests.' Sales to Acopio also influenced strategy. 'We changed from planting maize to sweetpotato because it needs less input and is higher volume – Acopio pays on volume,' explained one CCS farmer. Peri-urban farms were orienting to the stronger urban demand for horticultural crops.

Business as usual

Some more traditional and isolated CCS farmers in particular perceived little impact of the Special Period on their farming system.

Working with nature

The increased awareness of working with natural processes became an alternative management strategy, characterized by what farmers generally termed 'a strive for biological equilibrium'.

Source: Wright, 2005

Changes in overall management strategies

Combinations of specific coping strategies were identifiable on-farm, the majority of which were state-backed. These could be classified into distinct types, as described in Box 7.3.

Each farmer and cooperative had his or her own logic in terms of the impact of the Special Period and their resulting coping strategies. Each farmer had a unique strategy. The logical flow of strategy by two farmers, on CCS and one UBPC, are captured in the 'problem tree' diagrams of Figures 7.3 and 7.4 respectively. The 'problem' was the Special Period or lack of fuel.

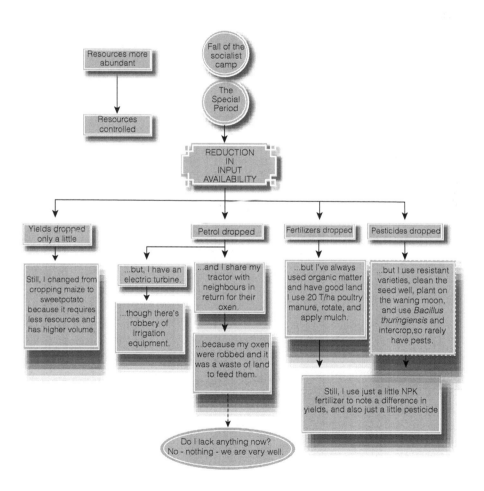

Figure 7.3 *Coping strategies of one CCS farmer during the Special Period: continuing with business as usual*

Source: Wright, 2005

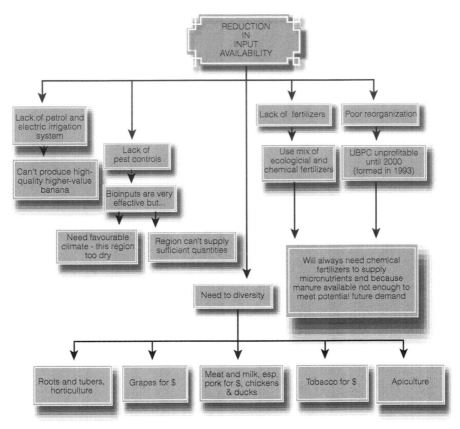

Figure 7.4 *Coping strategies of one UBPC farm during the Special Period: diversification*

Source: Wright, 2005

Farmers' experiences with farm reorganization and incentivization

Farmers felt that socio-economic factors were less of a constraint to production than technical ones. Their main socio-economic limitations were related to incentives, provision of information, opportunities to exchange and share with other farmers, and technical training. Figure 7.5 depicts farmers' perspectives on the perceived main socio-economic constraints to production.

Impacts of cooperativization and its support

All the cooperatives were associated with a specific municipal or provincial state Agricultural Enterprise, according to their specialized farming activity, such as livestock, citrus or mixed crops. This Enterprise assisted in the implementation of state production plans, and provided technical assistance and distributed resources. Several tensions arose in these relationships. For a start, the

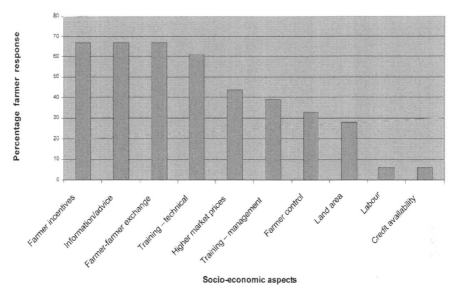

Figure 7.5 *Farmer identification of the main socio-economic constraints to production, 1999–2000 (% response from sample)*

Source: Wright, 2005

association by specialist commodity activity did not concur with the new drive for farm diversification; specialized Enterprises had difficulties in supporting the multiple activities being increasingly implemented on-farm. In one example, the strengthening of one CCS cooperative in Cienfuegos required it to become linked to the local state Enterprise for the first time. The Enterprise specialized in tobacco production, but the CCS members continued to produce mixed crops for which they actually received little support. Generally, UBPCs received greater support than the *campesino* sector, and CPAs more support than CCS cooperatives.

Configurations over land access were complex and flexible. Even CCS members, who owned land, were able to access more in perpetuity if they were able to demonstrate good performance results and agree to the production of specific crops. One CCS cooperative held almost half its land in this way. The state also rented land from private *campesino* land owners.

By the end of the decade, membership figures for both CCS and UBPC cooperatives were peaking. The strengthening of certain CCS cooperatives resulted in increases in membership as the benefits attracted the extended families of existing members. One CCS farmer explained: 'There is now a better income and other benefits such as having access to a shared tractor.' Similarly, on those UBPCs which had survived, labour demand had increased owing to the lack of machinery. Compared to state farms, UBPCs could, according to one member, 'guarantee work all year round and offer a better level of wellbeing'.[4]

Impacts of localized management of large-scale production

The UBPC programme of 'linking man to the land' was hugely popular and showing increasing benefits. Box 7.4 describes this system of operation on one UBPC farm in Havana Province.

Box 7.4 *Managing large-scale plantations on a human scale: linking man to the land*

The 63ha of banana plantation on one UBPC in Havana Province had been divided into 12 fields, each 5ha field having two linked workers (*vinculadores*). Each *vinculadore* tended 20 rows of banana, and could choose whether to work in pairs or alone. Every six fields had an area manager, and the whole group of linked banana workers came together over input applications and other issues. During the period of establishment of the plantation, the *vinculadores* earned $20/month for 'keeping their rows clean', and then an additional 0.5c for each healthy plant. If there was wind damage, they would have to replant, and were penalized 5c for each dead or neglected plant. The *vinculadores* also had to pay for any technical or input support required for their designated area. At the end of the period, the profit would be calculated and each worker would receive 50 per cent of the profits from his or her area. Salaries on this plantation were estimated at $40/month.

One *vinculadore* had just arrived on the UBPC six months previously and was now responsible for 2.7ha containing 8000 banana plants. The state Enterprise prepared the land for planting, by rotavating, levelling and removing stones. The *vinculadore* then undertook the planting – covering each banana microplant with 5cm of soil just to the level of the leaf. Then the UBPC's own plant health expert came in to undertake fertilizer and pest control applications, and other workers removed any suckers from the plants. An ox-handler, also linked to the banana plantation, visited as necessary. The *vinculadore* helped out with input applications and also hoed each day. He had never been on a training course, as he was, as he put it, 'just a banana worker. Some more able types may go on training courses.'

Source: Wright, 2005

Unlinked workers, who instead continued to be rotated for short periods over different crops and areas, were aware of the benefits of linking. 'If I got the opportunity, I'd be linked to banana production and sell it to Fruta Selecta. Then I'd work voluntarily, for longer hours. I'd take more care at harvest time, remove the weeds, do better soil preparation, remove the diseased leaves and suckers, remove the male buds, chop the leaves and put them between the plants to maintain humidity', expressed one unlinked UBPC member.

Changes in production costs, yields and incomes

With increases in farm responsibility came the increased awareness in financial management. Farmers' opinions varied as to whether production costs had risen or fallen over the decade. On the one hand, input and labour costs had increased, while, on the other, market prices and yields had increased and management improved. In Havana Province, near the capital, farmers were more market-oriented and production more intensive. Here, also, prices of organic inputs were more expensive.

Farmers agreed that farm yields had dropped in the early 1990s, by 15–50 per cent, but by the end of the decade yields were steadily increasing. Although 70 per cent of farmers attributed this decline to input availability, the remainder did not and identified a range of other influential factors. The weather and climate change was a major factor, as one CCS farmer explained: 'In the early 1990s there was only a small drop in yield. Now it's much worse because of the drought and the climate – there is less rain now and the sun is stronger.' In fact, a major factor influencing yields, pointed out by an organic CCS farmer in Holguin and concurred with across the country, was the cyclical agroclimatic pattern that resulted in very good yields every four years, specifically 1988, 1992, 1996 and 2000. This observation indicated that a long-term view was required when analysing yield fluctuations. Other farmers identified changes in support and institutional structures as affecting yields, such as one UBPC farmer: 'Over the last three years productivity has been improving. Yields are higher because more attention is paid and there are more incentives.' For some crops, such as potato and tobacco, yield figures quoted in 2000 were fairly standard across provinces and types of cooperative (ranging from 16.8 to 20.1t/ha for potato and 1.0 to 1.3t/ha for tobacco). For other crops, such as tomatoes, yields varied considerably depending on the conditions of each particular farm. In 1999, maize grain yields ranged from 0.1 to 1.8t/ha. This was felt to be a poor year, although for varying reasons: insufficient petrol in Cienfuegos, drought in Havana, and poor soils in Holguin. For banana crops, yields in the same period averaged 20.3t/ha for fruit and 11.2t/ha for plantain. Hurricanes had hit plantations particularly badly.

Market prices, though fairly similar across the provinces, varied between sales outlet and also changed dramatically by month, season and year. Taking the period 1999–2000 as an example, maize sold at 0.4c/cob or $24/t grain to Acopio, 1–5c/cob or $164/t grain in the farmers' market, and up to $330/t grain to the Seed Enterprise. Similarly, during this period, first-grade banana could sell to Acopio for up to $33/t, and second-grade for $11/t. Over-fulfilment of banana fetched $38/t from Acopio, between $27 and $77t at the farmers' market, or up to $33/t in the local controlled sales point Las Placitas. Frutas Selectas, when purchasing quality banana for the tourist trade, was paying $28/t.

Wages varied between and across all types of cooperative. Some CPA members in Havana province were taking home $23 per month including bonuses. Away from the capital, a typical CCS farmer might earn $40/month, although CCS farmers in general were the most wealthy of all cooperative types. A UBPC worker involved in a profit-sharing scheme might take home $20 to $60/month plus a bonus of $30 to $150/month, whereas those workers only on a basic wage might take home $11/month. Farmers producing an export crop such as tobacco or citrus would receive a stimulus in dollar currency, which could be in the region of $150 at harvest. Contracted labourers were being paid $1.50/day in Havana (plus food benefits), which was more than three times the wage of the 1980s. Generally, farmers agreed that salaries had risen considerably since the 1980s. On UBPCs, almost as important as salaries were the other benefits and welfare of the members, these being critical to maintain production performance. The economic balance record as kept by one CPA cooperative between 1982 and 1998 is given in Table 7.2. This shows that although yields fell after 1993 and were just beginning to recover, as did profit share, wages did not follow this trend but rose consistently.

Table 7.2 *Summary of production performance and staff income for one CPA, between 1983 and 1998*

Farm indicator	Year				
	1983	1988	1993	1997	1998
Total production t	2530	2642	3011	2236	2393
Yield t/ha	17.3	17.3	19.6	15.4	16.4
Annual basic salary $	32	46	54	68	64
Daily basic salary $	0.40	0.47	0.50	0.59	0.61
Member's profit $/day	0.40	0.40	0.69	0.42	0.39

Source: Wright, 2005

More diverse and flexible marketing arrangements

The state reforms in the range of market options had also had a positive impact on farmers. One CCS farmer affirmed: 'In the 1980s, all produce went to Acopio and the plan was more fixed because there was no other market, so the CCS did not have a marketing department then, and it received no extra payment for over-fulfilling the plan.' Farms could now deliver their produce to the following destinations:

- the state ration through the production plan (obligatory)
- self-provisioning of farm members, including workers' cafeteria
- farmers' markets and controlled sales points (for produce surplus to the plan)
- State Seed Enterprise (if producing for seed)
- state tourist market (through Frutas Selectas/Acopio)
- State Livestock Enterprise (for animal feed)

- social institutions (local hospitals, schools, old people's homes – semi-obligatory)
- the black market.

The changes in the way that the state production plans were developed had also made an impact. Plans were drawn up between the cooperative or farmer, Acopio and the associated state Enterprise or ANAP, to deliver an overall tonnage of produce per year. These were then broken down by crop, with agreements made to deliver a certain tonnage of the harvest or cropped area, and were renewed on a quarterly basis. So, for example, of the 13.4ha of plantain produced by one CSS in Holguin, 6.7ha was in the state plan to provide 36t banana (5.4t/ha). The same cooperative was also growing 54ha maize, all of which was dedicated to delivering 32t to the plan (0.59t/ha). Plans could be made over months or years: one CCS was providing 4.5t per month of banana from 2ha (2.3t/ha) over a period of five years, although its actual yields were more like 23.5t/ha. Figures indicated that yields agreed in the plan were not ambitious, and this allowed leeway for crop failures and also for enabling a surplus to go to other market channels. The majority of farmers felt that the quantities demanded by the plan had decreased since the 1980s, and this was largely because of the lack of inputs. The production agreement used to include provision of a technical package of inputs, but this was now less frequent and might only apply to those crops and producers prioritized by the state. In these prioritized cases, an annex was included in the contract stating that if the inputs did not arrive the quantities expected would be reduced to reflect this. In Holguin Province, one farmer explained: 'Last year I got no fertilizer from the CCS administration, and this year I got 0.1t. We have a contract and a plan, but the inputs never arrive.'

The prices paid by Acopio rose during the 1990s, and this, coupled with the introduction of purchasing surplus produce at a premium, aimed both to encourage production and to weaken the power of alternative markets. Yet Acopio also reduced its guaranteed purchases, and towards the end of the decade was only purchasing what was required for the ration. Its prices were more variable too: for example, in Holguin, its purchase price for maize had dropped by one-third over three years because, as one CCS farmer explained, 'Acopio has an abundance of maize now – their storehouse is full.' Farmers were unused to this partial removal of a guaranteed market. However, Acopio was also now more flexible, as one UBPC farmer from Cienfuegos explained: 'Acopio allows us to take what we've produced for the plan to Havana and sell it there at a higher price.' If state plans were under-fulfilled, then Acopio would levy a fine, unless the failure was due to factors beyond the farmer's control, such as bad weather or the state defaulting on the supply of agreed inputs. UBPC members might also receive disciplinary action and a reduction in wages. Farmers could and did take out insurance against this risk of under-fulfilment.

This insurance was costing one UBPC farmer $130 per year to cover 13ha of banana against pest, disease and inclement weather. The weather, and particularly drought, was usually to blame for under-fulfilment, while hard work and good state support were the reasons given for over-fulfilment.

Farmers' sales strategies were as diverse as their numbers and again demonstrate the uniqueness of each farm. Some cooperatives simply made agreements with Acopio for their whole crop, such as one CCS farmer who sold 30 per cent of his bananas at Acopio's basic price of $6.8/t and the remaining 70 per cent at the surplus rate of $17.6/t. 'Acopio pays double for over-fulfilment, and triple if the quality is good enough for Fruta Selecta,' explained another CCS member. Sales to other destinations varied. For one CCS in Cienfuegos, for example, 50 per cent of its maize harvest went to Acopio, 30 per cent for its own self-provisioning, and the remaining 20 per cent to the farmers' market. In contrast, a UBPC in Holguin was selling only 10 per cent of its banana crop to Acopio; the rest went to the workers' cafeteria (37 per cent), self-provisioning (21 per cent) and the farmers' market (32 per cent). Plans were not made for all crops grown, as explained by one UBPC farmer: 'we don't have a state plan for banana, so what we produce we use for self-provisioning and send some to the Tobacco Enterprise to which we are annexed. The rest we sell in the local Las Placitas market, to schools, to Acopio, and to private purchasers – these pay the best.' In fact, despite the diversity of markets, many farmers did not avail themselves of these. The reasons given related to costs, ease, labour, transport, market demand and marketing expertise. Several farmers preferred to sell to Acopio. 'There are too many middlemen and I would have to pay them all a salary and provide meals each day, and besides, the Acopio price is almost the same,' explained one UBPC farmer. Others took advantage of the farmers' markets, such as one farmer whose maize was fetching five times the Acopio price because it was out of season.

On-farm food self-provisioning

Farms were encouraged to renounce their rights to state rations when they became more self-sufficient. This showed that they were not a burden on the state, although some who had done this still received certain items such as sugar, oil and soap. The nature and type of self-provisioning varied by farm type. On some CPA cooperatives, self-provisioning took up a relatively small percentage of land area – only 8.3 per cent on one cooperative. In material terms, these CPA farmers were relatively poor. Individuals had incomes of less than one dollar a day, including profits. Yet living costs were low: housing was provided (though basic and often cramped), schooling was free, basic foodstuffs were supplied at a very low price, and medical care was free (although medicines were in short supply). Occasionally work clothes and boots were provided. Although protein consumption was generally low due to the limitations in livestock production and the small quantities of meat available through the ration system, each farmer member had his or her own family home garden where goats could be reared. The main food expenditures were on items unobtainable through the cooperative or ration system, such as cooking oil, and this competed with necessary household expenditure on clothing, use of public transport, and children's needs. Unless the cooperative had made an arrangement with the state to produce and sell export crops that would bring in a dollar income, food items that could only be purchased in dollars were inaccessible. There was little evidence of on-farm processing or value adding.

In the private CCS sector, much land was being given over to self-provisioning, and this brought with it the promotion of 'perennial production strategies' such as, for example, planting of early and late varieties of mango and cassava to extend seasonal availability. CCS farmers attempted to be individually self-sufficient, producing up to 90 per cent of their food needs. However, it was generally only the older farmers who actually achieved this; the younger, less experienced members had to resort to purchasing more food. CCS members could access dollar food stores, and estimated that the ration supplied approximately 8 per cent of their food needs, and dollar outlets 2 per cent.[5]

In the early days of the UBPCs, almost all produce was sent off-farm, with only the member's cafeteria being self-provisioned. Towards the end of the decade, this had changed, and produce was being sold to member families at cost price – largely grains, roots and tubers, and horticultural items. Some UBPCs provided more food, depending on their production performance. As an example, one UBPC in Holguin Province supplied its members with milk on a daily basis, four eggs per week, meat and other foodstuffs. In another, members were purchasing beans for $0.065/kg from the farm, compared with the price at the local farmers' market of $0.44–0.66/kg. Generally, UBPC workers had access to peso goods shops via their associated Agricultural Enterprise. These shops carried a wider range of products, and of higher quality, compared to those found in dollar shops, and at a subsidized price. Some UBPC members also produced their own extra foods, working a small plot of land or *navé*. This land was rented to the members by the cooperative, payment being made in-kind as a share of the harvest. UBPCs produced between 65 and 90 per cent of their food needs on-farm, and members felt that the ration supplied another 15–25 per cent, while 10 per cent came from workers' self-provisioning plots.

Overall, and compared to the levels of insecurity over food availability amongst the population in general, farmers were wholly positive about the increases in domestic production that they had experienced since 1995, and by the end of the decade they felt that there was more food available. That the farming sector felt more food secure than the general population could be attributed to their direct control of their own food supplies, particularly through their improved self-provisioning status.

The significance of continued petrol and agrochemical use

Petrol and agrochemical input distribution and usage was uneven between regions, farms and crops, and this created a national patchwork effect of more and less intensive production. Although this pattern was caused largely by resource constraints, it was also affected by production plans, by urban and peri-urban organic production, by areas where a local variety was particularly robust or a particular organic input easily available, or by innovative individuals and institutional programmes. For example, Havana Province was more intensively farmed than its neighbour Pinar del Rio, sugar received more agrochemicals than maize as did seed crops destined for the Seed Enterprise, and high-performing farmers received more inputs again.

Overall, and based on the continued use of petrochemicals and the intention to use them more widely when possible, there were no instances of a rural farm proactively practising organic agriculture across the whole entity. The prioritization system systematically valued agrochemical farming and undervalued organic farming. Yet in the absence of sufficient on-farm vision, knowledge, capacity and motivation to properly implement viable organic systems, this system proved a successful strategy for maintaining production levels during critical years, not only in terms of its physical contribution to yield performance, but also in terms of maintaining a degree of moral cohesion in the agricultural sector. The successes of higher-potential farms served as models for other farmers, and morale and hope were sustained. That some petrol was still available, however slight, indicated to farmers that the situation had not completely fallen apart and was going to get better. At the same time, farmers were building their knowledge on alternatives. These changes in the farming sector indicate the huge amount of learning that had taken place, both official training and spontaneous experiential learning. Although this learning did not show greater productivity gains in the short period up to the end of the 1990s, its benefits were likely to emerge over the longer term. Cuban farmers agreed that they would not go back to the high-input ways of the 1980s, and they were more prepared for a longer-term fossil energy descent.

Although farmers felt that petrol was the main limiting factor to production, when asked how they had achieved higher yields, they all identified the changes in land tenureship and in farm organization as being more important contributory factors than input availability. These changes had resulted in greater farm efficiency and higher levels of care and attention, and came with more support from the state. This perspective was substantiated by the increasing dominance of the *campesino* sector, which by 1998 was producing 86 per cent of tobacco, 68 per cent of maize, 73 per cent of beans and 47 per cent of roots and tubers (ONE, 2000).

Notes

1 Herrera Sorzano (1999) identifies an important difference between cooperatives in other countries and those in Cuba. The former have focused on developing secondary activities such as processing and marketing, whereas in Cuba the focus remains on primary production. This may partly explain why Cuba's cooperatives encountered difficulties in becoming economically efficient.

2 *Bacillus thuringiensis* is a bacterium that parasitizes the caterpillars of some harmful moths and butterflies. Spraying or dusting plants with spores of this bacterium is accepted as an environmentally safe way to control such pests.

3 Contemporary scientific data from Cuba supported this. In a survey of 450 *campesino* farmers in nine municipalities in the central provinces of Villa Clara, Sancti Spiritus and Cienfuegos, Rojas et al (2000) found that 54.5 per cent of farmers were using only chemical pesticide control for maize leaf borer, the most common being a mixture of Parathion and Carbonil. Biological controls, meanwhile, were being used by 3.3 per cent of *campesinos*, mainly *B. thuringiensis.* Only 1.3 per cent were using both types of control. Of the farmers who used chemicals, 50.3 per cent did so because they felt them to be most effective, whilst 33.6 per cent did so because they had access to them. Of farmers who used biological controls, 76.4 per cent did so because they were beneficial, 12.5 per cent because of ease of access, and 11.1 per cent because of low costs. Chemical pesticides would be applied just once, and biological on a weekly basis.

4 Martín (2002) identifies specific changes in the agricultural labour force as a result of the Special Period and reforms: these include an increase in internal heterogeneity as a result of the emergence of new cooperatives and joint ventures, an increase in the size and diversity of the labour force, increased numbers and economic importance of *campesinos*, increased numbers of cooperative members, reduction in numbers of less skilled farm workers, and greater differentiation in incomes and living standards as a result of market linkages and technologies.

5 Other evidence supports this situation of CCS cooperatives. Ríos Labrada et al (2000) found that CCS farmers in Pinar del Rio Province produced tobacco as their only marketable crop. They self-provisioned a wide variety of other foodstuffs, ranging from staples to condiments, which were all considered too expensive to purchase, and also undertook home-processing.

References

AGROINFOR (1998) *Actualidades de la Agricultura*, Boletín 0/98, Havana, MINAG

CNSV (2000) *Estadisticas Centro Nacional de Sanidad Vegetal*, Havana, MINAG

Deere, C. D. (1996) *The Evolution of Cuba's Agricultural Sector: Debates, Controversies and Research Issues.* International Working Paper Series IW96-3, Gainesville FL, Food and Resource Economics Department, Institute of Food and Agricultural Sciences

Deere, C. D. (1997) 'Reforming Cuban Agriculture', *Development and Change*, vol 28, pp649–669

Deere, C. D., Pérez, N. and Gonzales, E. (1994) 'The view from below: Cuban agriculture in the "Special Period in Peacetime"', *The Journal of Peasant Studies*, vol 21, no 2 (January), pp194–234

Díaz, B. (1995) *Biotecnología Agrícola: Estudio de Caso en Cuba*, Washington DC, Latin American Studies Association

EIA (2008) *Cuba Energy Profile*, Energy Information Administration. http://tonto.eia.doe. gov/country accessed in March 2008

EIU (1997) *Country Report: Cuba. 4th quarter 1997*, London, The Economist Intelligence Unit

Enríquez, L. J. (2000) *Cuba's New Agricultural Revolution. The Transformation of Food Crop Production in Contemporary Cuba*. Development Report No. 14, Dept. Sociology, University of California. www.foodfirst.org/pubs/ devreps/dr14.html

Everleny, O. and Marquetti, H. (1995) 'Comportamiento de la economía cubana en 1994. Tendencias', *Boletín informativo CIEM*, no 21, pp3–15

FAO (2003) *Fertiliser Use by Crop in Cuba*, Rome, Land and Water Department Division, FAO

FAO (2004) *Food and Agriculture Indicators. Cuba*. www.fao.org/countryprofiles accessed in 2004

Fernández, P. (1998) *Cuba Toward the Third Millenium*. Seminar by Pablo Fernández (MINAG, Cuba), Wageningen University, 9 October

Granma (2003) Speech by President Fidel Castro (29 December)

Herrera Sorzano, A. (1999) *Cooperativisimo Como Forma de Tenencia y su Papel en la Organización del Espacio Rural Cubano*. Doctoral thesis, Faculty of Geography, University of Habana

Lage, C. (1995) 'La economía cubana en 1994', *Boletín informativo CIEM*, no 19, pp187–200

Martín, L. (2002) 'Transforming the Cuban countryside: property, markets and technological change', in Funes, F., García, L., Bourque, M., Pérez, N. and Rosset, P. (eds) *Sustainable Agriculture and Resistance: Transforming Food Production in Cuba*, Oakland CA, Food First Books, pp57–71

Miedema, J. and Trinks, M. (1998) *Cuba and Ecological Agriculture. An Analysis of the Conversion of the Cuban Agriculture towards a more Sustainable Agriculture*. Unpublished MSc thesis, Wageningen Agricultural University

MINAG (1998) *Movimiento Cooperativo y Campesino. Indicadores Seleccionados*, la Havana, Direccion de Funcionamiento y Desarrollo Cooperativo y Campesino, Area de Recursos Humanos, MINAG (May)

ONE (1997) *Estadisticas agropecuarias. Indicadores sociales y demográficos de Cuba*, Havana, Oficina Nacional de Estadísticas

ONE (2000) *Anuario Estadistico de Cuba 2000*. La Habana, Oficinal Nacional de Estadisticas.

ONE (2006) *Anuario Estadístico de Cuba 2006*. Edicion 2007, La Habana, Oficina Nacional de Estadisticas

Paneque Brizuelas, A. (1997) 'Food for thought. Agriculture's outlook at year-end', *Granma International 1997*, Electronic Edition, Havana. www.granma.cu/1998/ 98ene1/50dic51.html

Ríos, A. (2002) 'Mechanisation, animal traction and sustainable agriculture', in Funes, F., García, L., Bourque, M., Pérez, N. and Rosset, P. (eds) *Sustainable Agriculture and*

Resistance: Transforming Food Production in Cuba, Oakland CA, Food First Books, pp155–163

Ríos, A. and Aguerreberre, S. (1998) *La Tracción Animal en* Cuba, Havana, Evento Internacional de Agroingeniería

Ríos Labrada, H., Almekinders, C., Verde, G., Ortiz, R. and Lafont, P. R. (2000) *Informal Sector saves Variability and Yields in Maize. The Experience of Cuba* (unpublished manuscript)

Rojas, J. A., Gómez, J., Morales, L., Sánchez, A. and Méndez, Y. (2000) 'Uso de la lucha biológica en el control de Spodoptera frugiperda (J. E. Smith) en el sector campesino de dos municipios de la Provincia de Villa Clara', *Proceedings, IV Symposium on Sustainable Agriculture of the XII Scientific Seminar of INCA*, 14–17 November 2000, Havana, INCA

Rosset, P. M. (1996) 'Cuba: alternative agriculture during crisis', in Thrupp, L. A. (ed) *New Partnerships for Sustainable Agriculture*, Washington DC, World Resources Institute, pp64–74

Rosset, P. (1998) *Eight Myths about Technology and Agricultural Development.* Edited notes from presentation given at the *Sustainable Agriculture Forum* (SAF), Vientiane, LAO PDR, NCA, 18 June

Rosset, P. (2002) 'Lessons of Cuban resistance', in Funes, F., García, L., Bourque, M., Pérez, N. and Rosset, P. (eds) *Sustainable Agriculture and Resistance: Transforming Food Production in Cuba*, Oakland CA, Food First Books, ppxiv–xx

Rosset, P. and Moore, M. (1997) 'Food security and local production of biopesticides in Cuba', *Leisa Magazine*, vol 13, no 4 (December), pp18–19

Sinclair, M. and Thompson, M. (2001) *Cuba Going Against the Grain: Agricultural Crisis and Transformation,* Oxfam America. http://www. oxfamamerica.org/newsandpublications/publications/research_reports/art1164.html accessed May 2002

Wright, J. (2005) *Falta Petroleo. Perspectives on the Emergence of a More Ecological Farming and Food System in Post-Crisis Cuba.* Doctoral thesis, Wageningen University

Wroe, A. (1996) 'Oranges and lemons: the limits of farm reform. Heroic illusions. A survey of Cuba', *The Economist* (6 April), pp8–10

8

Institutional Coping Strategies: Transition and Decentralization

In Cuba more than most other countries, the farming system is both planned and managed by the state. This means that changes in farming practices and processes were both partially determined by, and would affect, state institutions, which would themselves be impacted by the conditions of the Special Period. As Sinclair and Thompson (2001) explain:

> Rural life in Cuba is not the litany of misery it is for *campesinos* in so many Latin American countries. Cuban farmers can count on rural institutions designed to support them, free, widely accessible social services, an impressive physical infrastructure, access to land, some inputs and a stable market and strong national cohesion in terms of social values.

These agricultural institutions had to deliver the new state reforms whilst at the same time respond to farmers' new demands and undergo internal changes themselves in order to survive. Although without private financial interest, agricultural institutions held substantial power and control, and some of this had to be relinquished.

Wider influences on change

Financial restrictions resulted in a widespread lack of resources that affected the whole support sector, from policy and administration to research and extension. This forced greater financial accountability and economic self-sufficiency across the board, as well as organizational restructuring and some degree of decentralization. These changes were both helped and hindered by the socio-political framework: helped, for example, by the horizontal and vertical fluidity of personnel who were obliged to rotate in and out of management positions within and between institutions,[1] by the high levels of education and by collective approaches; hindered, for example, by the continued poor access to international research journals, the travel restrictions, and the sectoral focus on technical agronomic issues with little attention paid to socio-economic aspects or evaluative studies. Foreign influences also played an important role: international private sector linkages provided funding for industrialized agricultural development, and NGO linkages for organic development projects.

By the end of the 1990s, the Cuban agricultural support sector was still in the process of transformation. Formal and informal institutional coping strategies varied from continuing with business as usual and attributing failure to externalities, to embracing completely new methodological and technical approaches. Although new approaches were often influenced, or even driven, by foreign interests, some were initiated internally, out of the experiences of the material limitations of the early 1990s or based on earlier pioneering work from the 1980s. Overall, a set of structures and processes were emerging to characterize the way that institutions were supporting food production.

The policy dilemma: increased yields or longer-term sustainability?

The main aims of Cuban agriculture in the 1990s – to increase yields and increase sustainability – were somewhat at odds with each other. The state had been planning some sustainable reorientation of production prior to 1989, with, for example, the establishment of the National Food and Nutrition Programme and the experimentation with farmers' markets. However, it was only after the depletion of buffering reserves in the early 1990s that major action was taken. Rather than a shift in collective or political attitude or proactive policy towards sustainability, towards increasing farmer involvement in decision-making or towards using more holistic research and development methodologies, it was the lack of petroleum and the enforced drive for self-sufficiency that acted as the gelling agents to stimulate a concerted move towards localization, diversification, resource streamlining and import substitution that occurred throughout the 1990s.

Policy development on environmental sustainability emanated from the Ministry of Science, Technology and Environment (CITMA). For example, Environmental Law 1997, Section 9, Article 132 was devoted to sustainable agriculture and called for a more rationalized use of locally appropriate industrialized and organic inputs and IPM measures (Pérez and Vázquez, 2002, pp136–137). However, MINAG, which was in charge of production and management of resources, had a singular mandate to increase production and did not have the resources to police the upholding of the law.

At the same time, although the term 'sustainable' was used in policy, there was little explicit, shared definition, and institutes had divergent strategies. For example, The National Nutritional Action Plan specified the desired strategy of 'sustainable development combining alternative with industrialized models:

including agroforestry, low input approaches, yet with a focus on intensification' (p68), and rather ambiguously went on to conclude: 'Given the conditions of the country, it should be maintained wherever possible, the application of high external and internal inputs on the basis of intensive practices, in others alternative practices, and ultimately an appropriate combination of the two models.' Meanwhile, the new Agricultural Extension System had 'an overall aim for high yields and low inputs'. MINAG was planning to increase the availability and use of agrochemicals, and its long-term (ten-year) crop production plans aimed for 60 per cent of production to be intensive, on 40 per cent of the land, and the remainder extensive. The ten-year plan for maize production, a relatively low-input crop, is given in Box 8.1.

Box 8.1 *The ten-year plan for maize production of MINAG, at 1999*

The objective of the plan was to obtain a national average yield of 3t/ha so as to supply the population with the national demand of 50kg/capita/year, plus an additional 300,000t of dry grain for livestock feed. The plan envisaged a three-pronged strategy:

1   'Intensive production': 29 per cent of total maize area sown under high input conditions using 100 per cent hybrids to achieve yields of 4.5t/ha and production of 390,000t, for human and animal consumption.
2   'Sustainable production': 47 per cent of total maize area under sustainable conditions using 70 per cent hybrids and improved varieties, to achieve yields of 3t/ha and production of 470,000t, about half of which was tender maize (premature), largely for human consumption
3   'Self-provisioning consumption': 22 per cent of total maize area under low-input conditions using 20–30 per cent improved varieties for yields of 1.5t/ha. This included the use of organic inputs only and with no mechanization or irrigation, focusing on marginal soils.

This plan required the tripling of maize cropping area and required the approval of MINAG to utilize land currently being taken out of sugarcane production.

Source: Wright, 2005

The 17 national agricultural research centres had reoriented their programmes toward sustainability, which entailed the promotion of three strategies: biotechnology to obtain high yields; integrated agriculture; and the combined use of biological inputs and agrochemicals (Díaz, 2000). The term 'low input' was ill-defined; one researcher attempted to explain it thus" 'Low input can be without any or with low amounts of inputs. Low inputs means low yields. It is substitution and adding manure. It also means being more precise – to measure exactly what the plant's needs are.' Nevertheless, research models that were referred to as 'low input' or 'integrated' were frequently in complete absence of agro-chemicals (e.g. Castro et al, 2000; López et al, 2000).

Within this, practical initiatives were diverse and promoted different forms of agriculture, yet on balance priority remained on increasing yields and increasing production. Summing up the tension between productivity and sustainability, one researcher described the results of some recent soil fertility trials:

> We recently ran trials of a new agrochemical compound fertilizer for potato, which came from Mexico. The formula contains specific minerals, and we found that potatoes yielded very well with this. But to achieve such high yields we exceeded the sustainable limit in terms of dosage, and the potato harvest contained over the legal limit of nitrate levels. Production costs were also higher. But this country's policy is to obtain the highest yields whatever the costs, and we in the research sector are not in charge of what happens with our results. I had wanted to use green manure in the trial as a sustainable comparison, but we were only given short notice to undertake these trials and had no seed available.

Factors forcing institutional change

Shortages of resources

Agricultural institutions were restricted in the type of work that could continue or be initiated. The shortage of fuel constrained their ability to reach field sites or participate in national meetings, and a shortage of other inputs meant that experiments, trials and extension work that were dependent on intensive approaches could no longer be pursued. The national research centres, based in Havana, limited their field work to within the province. Institutional gene banks lost a lot of planting materials; one estimate gave a loss of 36 per cent of pasture, 24 per cent of citrus and other fruit, and 14 per cent of rice planting material. Equally, there were fewer resources such as paper and printing equipment available to publish or exchange research findings; in fact, many findings from this period continue to lie in draft form in desk drawers. The inability to print meant that professionals only had recourse to outdated information produced before the crisis. In one agricultural library, for example, of the 37 Cuban scientific journals in stock, only 13 contained volumes dating after 1993, and only four of these were continuous series. Foreign journals petered out completely after 1990. Organic initiatives were also disrupted by these changes; for example, the development of bio-fertilizers was restricted by a lack of substrate material. This inability to continue with business as usual meant that institutions had to change, and, for a start, staff were obliged to carry out extra tasks such as food self-provisioning and security work.

The enforcement of economic responsibility

As well as resource restrictions, subsidies to the agricultural sector were significantly reduced. Institutions within MINAG had to generate at least some of their income, and this was termed the 'enforcement of economic responsibility'. After 1997 the sector had to be self-sufficient for all its dollar requirements, which for some institutes included electricity usage. These restrictions led to an increase in institutional accountability and a revaluing of work so that some activities felt to be inapplicable to the conditions of the Special Period were wound down.

Many research institutes were attempting to commercialize products, such as machinery or bio-fertilizers (based on mycorrhiza, Rhizobium and neem compounds), and services, although most did not have the relevant training in business management or marketing. MINAG helped out by purchasing these products in dollars and selling onto farmers in Cuban pesos. Universities introduced fees for research services even though these fees covered only 10 per cent of their real costs. Some Agricultural Enterprises extended their remit for self-provisioning by producing crops for sale, and *campesino* farmers were charged for inputs – though at 50 per cent of their real costs – and for some research services. MINAG paid for services to state farms.

Many institutions struggled with this change as they lacked the know-how and resources to initiate successful responses. Equally the economic situation was not favourable for setting up new initiatives, as everybody's purchasing power was diminished. In this scenario, institutions looked abroad for financing, for developing agribusiness partnerships and development aid projects. Even here, change was required. Whereas previously institutes had primarily interchanged with socialist countries, now they were forced to seek and forge new and stronger partnerships with institutes in other regions, such as Europe, Canada and South America.

The economic situation was not helped by the fact that institutional salaries were low. In one provincial research centre, for example, a chauffeur would receive $7.5/month, and researchers received $12.5–$15/month, with a maximum of $25 depending on their scale. The director was remunerated with an additional $0.75/month on top of his or her salary scale.

Institutional and individual coping strategies

Institutions now had to support the changes and reforms on-farm: the strengthening of the *campesino* sector, the development of UBPCs and the greater financial responsibility of ex-state cooperatives, the higher demand for organic inputs, the development of *organoponicos* in towns and on farms, the general downscaling and diversification, and so on. It was, in short, an unsettling period, with professionals having to take on new responsibilities, often outside their disciplines and training. Yet it was one in which new opportunities also emerged that some were able to take advantage of. Various coping strategies emerged, manifesting through explicit changes in institutional policy and strategy as well as through an implicit repositioning of behaviour and attitude. These strategies are described as follows:

- Business as usual: some institutions, departments and individuals carried on using the same approaches as previously, and blaming externalities (such as the weather, sanctions, farmers or policy) when these approaches did not work. They were waiting for external conditions to change.
- Focus on the minority and ignore the rest: some recognized that external conditions were different but still chose to focus on those areas and issues that had remained the same or that were low risk, such as supporting well-resourced farmers or existing state farms.
- Institutional bravado: some were adamant that they had not experienced difficulties in the Special Period, and whilst this was at least partly a brave face put on for a foreigner, it did indicate that they were not coping very well.
- Try something different: many individuals and institutions were to some extent taking up the challenge to change and seek alternatives, be it different methodological approaches or promoting the new diversification options of Raul Castro.
- Opportunity to flourish: those who had previously been working on unpopular or maverick-type issues found that they suddenly had something useful to contribute, and used this opportunity to take their interests into the mainstream.

Maintaining state planning, encouraging local management

Partial decentralization and streamlining

Institutions in the agricultural sector were located under the Ministry of Agriculture (MINAG), the Ministry of Sugar (MINAZ), the Ministry of Science, Technology and the Environment (CITMA), and the Ministry of Higher Education (MES). The basic infrastructure of the agriculture and food-related institutes as it stood in the early 1990s is shown in Appendix B. MINAG comprised delegations in all provinces and municipalities of the country, as well as research and extension institutes, enterprises, input producers and suppliers, Acopio, and more. However, the headquarters of all the major institutes were located in the capital. As ministry staff explained: 'Almost all institutes were focused in Havana, so interaction was all vertical, each institute with its own approach and duplicating work.' A period of decentralization began in 1994, and by 1998 this resulted in a dramatic downsizing of MINAG, starting from the top. Table 8.1 shows the decrease in staff in the Ministry's Havana headquarters.

Table 8.1 *Changes in staffing and structure of MINAG HQ, comparing 1990 with 1998*

HQ staffing and structures	Nos. in 1990	Nos. in 1998
Minister	1	1
Vice-minister	10	8
Directorates	38	25
Departments	104	24
Sections	12	0
Employees	1350	252

Source: Miedema and Trinks, 1998, from MINAG, 1998

The National Food Plan became subsidiarized to provincial- and municipal-level offices. However, the state maintained its centralized planning and control over production plans and food security issues. This meant that the sector could rapidly implement new policy and effectively react to threats and opportunities, such as the nationwide outbreak of the *Thrips palmi* pest, which was brought under control within three years.

Other areas of government decentralization were around resource management, market decision-making, and some aspects of research and extension approaches. In attempts to streamline and increase efficiency, some functions were shifted from one state entity to another, such as handing over the management of domestic tourist food supply to the Citrus Corporation. At the end of the decade, moves were finally made to sort out the functioning of the state Seed Enterprise and improve seed provision. The Seed Enterprise had historically produced and disseminated a small range of varieties for each crop (12 in the case of maize), these varieties requiring medium-high levels of inputs. The Enterprise had previously relied on state farms to multiply seeds under high input conditions. By the early 1990s, this supply had fallen to 50 per cent of previous levels. Nevertheless, at the end of the decade, the Enterprise was still recommending monocultures and agrochemicals. As one staff member explained, 'We have not been interested in low input seeds, and, anyway, quality is the most important aspect, such as germination and phytosanitary conditions.' The Enterprise was also unenthusiastic about farmer seed-saving. 'The producer always mixes varieties and may harvest when it is humid or do other things which lower seed quality,' explained one Enterprise staff. Another, although recognizing that state supplies had fallen, felt that 'The situation is now stabilizing and producers do not need to save their own seed any more.' Still, in 1999, responsibility for seed certification passed from the Seed Enterprise to the territorial stations of Sanidad Vegetal, and for supplying seed to the Agricultural Enterprises at municipal level. Seed multiplication was transferred from state farms to the more productive private farming sector. This did not prevent Seed Enterprise staff from maintaining an industrialized perspective on the future; in order to solve their problems, they saw the way forward as being through joint venture schemes with foreign companies, to maintain large-scale production and using high levels of technology, 'such as aeroplanes to spread the herbicide'.

Concurrent with this, some centralization continued to occur. For example, MINAG took more control over the grassroots organic movement, ACAO, and by strengthening CCS cooperatives also achieved some control over these *campesino* farmers. In many respects, the paternalistic approach of the state to farmers continued. This was to some degree positive: the farming sector remained subsidised to some degree, and institutions providing services, research and extension were maintained with high and qualified staffing levels. Towards the end of the decade, Cuba had 221 research and development centres and 46 centres of higher education, and employed over 60,000 workers (CITMA, 2000). The downside was that institutional staff such as extensionists still maintained a strong influence over production strategies; their views were interpreted as prescriptive and indicative of state policy, especially amongst ex-state farm workers who were unaccustomed to making their own decisions over production and management. This then led to institutional hesitancy over the extent of farmers' capacity and the degree to which they could be given autonomy over activities such as the on-farm production of bio-fertilizers and seed saving. Even in the face of emerging evidence that farmers were capable of self-management, the majority of institutions maintained control of these activities.

At the same time, some measures were taken to encourage interaction with farmers as stakeholders. The centralized production planning system already provided some avenues for interaction and feedback; at local level this was through ongoing planning agreements, and at national level a selected group of *campesino* farmers acted as advisors to MINAG's Council of Ministers. Increasingly, technical bulletins on production were drawn up in consultation with farmer groups at municipal levels. Farmers' willingness to participate varied. *Campesino* farmers were reticent to divulge information to the state over their actual yields and incomes, in case this obligated them to sell more to

Acopio, yet where they were given autonomy and responsibility they became more interested in participating in meetings and events.

Attempts at integrating support for farmers

With the changes in farm structures, so the decade saw attempts to integrate and increase the efficiency of the entities that directly supported them. The Agricultural Enterprises had already been partly merged with Acopio in 1995 to provide a 'one-stop shop' for marketing, production and service provision to the UBPCs. In 1999, the Enterprises were again overhauled in a remodelling based on the structure of the longstanding and successful commodity corporations. This overhauling resulted in the formation of 25 national Economic Organizations responsible for the production of mixed crops, tobacco, citrus, livestock, coffee and so on. At provincial level, the Agricultural Enterprises were tasked with providing training and services. They purchased inputs from a central supply and distributed them to municipal-level Enterprises for provision to farmers. At the same time, they were responsible for encouraging diversification and developing and testing new technologies with selected farmer leaders. In terms of training, they ran courses on business skills, and some more wealthy Enterprises were developing business service centres for farmers, offering software, photocopying, training courses, email and accounting. This, along with the hiring out of their assets such as storehouses and workshops to farmers, helped the Enterprises to become more financially self-sufficient.

Concurrent with the changes in the Enterprises, in 1995 the Campesino Cooperative Sector (SCC) lost its historical role of providing inputs and resources to the *campesino* sector when these tasks were taken up by the Enterprises. In 1996–97, the SCC was then merged with provincial level Enterprises that had previously only served state farms and UBPCs, and ANAP was also implicated in these changes; it lost its role in providing agricultural support and had to refocus on socio-political tasks. These changes affected the *campesino* sector badly. A staff member of ANAP explained: 'The *campesinos* lost confidence in ANAP because it could no longer supply their inputs and so on. They were used to going to ANAP to talk about their problems. They still do this.' The SCC had, along with ANAP, developed a close network with its farmers, with representatives visiting on a regular basis to solve problems and provide inputs including work clothing. Importantly, the SCC had had sufficient resources to continue supplying agrochemicals up until 1996. The organizational merger coincided with these supplies running out, and the Enterprises did not have the resources or motivation to attend to the *campesinos* to the same degree. Further, the Enterprises were specialized by nature and unsuitable for supporting the diversity of *campesino* production. As one provincial Enterprise staff explained, 'The problem is that there are no pure producers

– they all have mixed crops.' In 1999, MINAG recognized the extent of the problem by bringing in a new resolution to partially reinstate the SCC and instruct the Enterprises to integrate *campesino* farmers into their structures, by supplying them with inputs and working together on production plans.

Dealing with the dealing out of resources

Breaking the equality

The crisis and the ensuing scarcity of resources had, as one professional put it, 'broken the equality', as inputs that had previously been available for everyone had to be prioritized. By the end of the decade, all inputs – including agro-chemicals, hoses, organic matter and biological inputs – were being distributed through the Agricultural Enterprises, and decisions on distribution were taken at provincial level, by Sanidad Vegetal, the Soils Institute, CITMA, universities and regional research centres. Each municipality had different priority crops, and different quantities of petrol dedicated to each. In October 2000, the following crops were prioritized in one municipality of Holguin Province: all potato, 40ha sweetpotato, 416ha cassava, 40ha banana, and also garlic, pumpkin and tomato. Inputs were designated for the farmers with the highest yields and productivity, whether CCS, CPA or UBPC. As one Enterprise staff explained, 'I only receive a few resources to distribute, so if a producer has water it is better to give him the inputs because there is more security that he will obtain good results.' Another justified this: 'Poor producers have to improve their techniques and they receive as much help in this as the others.' Agrochemicals were sold to farmers in pesos, whereas petrol required dollars, and, notwithstanding the careful distribution, there were indications that farmers could access more inputs fairly liberally if they could afford it.

Integrated input recommendations

By the end of the decade, the Agricultural Enterprises were advocating the integration of chemical and biological inputs. MINAG's technical recommendations for roots and tubers crops, for example, stated, 'An attempt has been made to recommend a fertilization system which is most practical and as adaptable as possible to the real conditions, with an emphasis on minimum quantities of fertilizer to obtain sustainable and acceptable yields,' and further, 'Chemical control should only be used when biological control has not been effective or for a very severe attack which is certain to severely reduce yields. As far as possible, use products selectively which have little or no lethal effect on natural enemies. In the case of applying insecticide, it will be necessary to re-establish beneficial fauna through releasing of biological controls' (MINAG, 1998). Table 8.2 provides an excerpt from the technical instruction leaflet of input recommendations for sweetpotato.

Table 8.2 *Input recommendations for sweetpotato from technical instruction leaflet provided by MINAG, 1998*

Crop	Management of seed	Management of soil fertility	Management of pest control
Sweetpotato	See instructions for establishing own seed bank	NPK at 0.4–0.6t/ha 1–1.5lb/plant of organic matter Bio-fertilizers (mycorrhiza, *Azobacter*, *Pseudomonas*)	Integrated control of weevil through: Use healthy seed and disinfect with *Beauveria* Apply *Beauveria bassiana* Use pheromone traps Use ant predators Irrigation as control Avoid intercropping Rotate crops Eliminate host plants Harvest on time Eliminate harvest wastes For red spider mite, apply Bifenthrin 1l/ha and Malathion 1l/ha% For Coleoptera, apply Carbaryl 2kg/ha and Malathion 1l/ha

Source: MINAG, 1998

That more agrochemicals were available was noted by one Enterprise staff: 'Previously we had to rely on biological inputs, but now we can apply a balanced mixture.' Recommendations from research institutes were more specialized and technically focused, based around specific varieties and hybrids and technology packages. Compared to recommendations of the 1980s, they nevertheless demonstrated a reduction in agrochemical instruction and an increase in alternative input suggestions. Strategies varied, however; one group of plant breeders, acknowledging the high costs of agrochemicals, were breeding for yields of 30–50 per cent higher than the norm in order for their new varieties to be cost-effective for farmers. Table 8.3 provides an example of input recommendations being promoted for commercial maize varieties and hybrids, these recommendations coming in the form of instruction leaflets in similar style to those for industrialized packages.

Table 8.3 *Input recommendations for commercial maize varieties and hybrids, from technical instruction leaflets of agricultural research institutes*

Maize Type	Potential yield	Fertilization	Input recommendations	
			Pest control	Weed control
Variety X	5.4–6.0t/ha For large-scale production with high or low inputs	N – 135kg/ha P – 80kg/ha K – 90kg/ha	Follow recommend-ations of Sanidad Vegetal	Apply one herbicide only if soil is com-pacted, and another after the cultivator has hilled-up
Double hybrid Y	10.5t/ha	N – 150kg/ha P – 90kg/ha K – 80kg/ha Depending on soil type. Apply when soil is humid and may cover	If Leaf-hopper: Dimicron 1l/ha Bi 58 1l/ha If Leaf-borer: Tamaron 1l/ha If Leaf-borer: Thiodan Parathion 30kg/ha Dipterex 30kg/ha	3–4 kg/ha Gesaprim Where this is not possible, weeds can be controlled by mechanical and manual means

Source: Wright, 2005

Provincial and municipal recycling and production of bio-fertilizers

Whereas agrochemicals were rationed and distributed to the farm, organic inputs could be obtained at will, where available, but had to be sought out. Some institutes and regions were supporting farmers to access organic inputs more than others. Agricultural Enterprises in one municipality in Havana, for example, were assisting in the recycling of farm byproducts: purchasing manure and maize husks from specialized livestock or mixed crop farms and transport-ing them to where they would be returned to the land or for use as livestock feed. In this case, they were supporting the local integration of crops and live-stock between rather than on-farm. Another Enterprise was producing its own compost from wastes collected throughout the region, and selling this to farm-ers. Yet another, at provincial level, was producing *bioterra* – composted sugarcane residue – and providing it free for collection by municipal Enterprises. In banana-growing regions in the east of the country, state farms and UBPCs were prohibited from removing fallen leaves and stems from the plantations as it was recognized that these contained high levels of vital nutrients, including potassium.

The use of micro-organisms for improving soil fertility had fallen since the 1980s, but was rising again towards the end of the decade; Rhizobium, for example, increased from 8300t in 1993 to 11,500t in 1998, and total bio-fertilizer production amounted to 2 million tons (Sinclair and Thompson, 2001; FAO, 2003). The production of worm humus and compost had commenced in 1992 and rose to 78,000t humus and 701,000t compost by 1994. After this it declined to a total production of 600,000t in 1998 (Treto et al, 2002). One worm humus production centre that had seen production drop by 50 per cent had been dependent on cattle manure as food for its Californian redworms. Over the decade, cattle were increasingly grazed outside and so manure was more hard to come by, so it was having to be substituted with filter cake from the sugarcane industry.

The network of provincial Processing Centres for Organic Matter and Vermiculture was small; the majority of biological soil inputs were produced by research institutes as an income-generating strategy, but these centres had no training in marketing and no transport for distribution. Staff at one research institute pertaining to CITMA described their difficulties in marketing their mycorrhizal product: 'There is foreign interest in our product, but we are less certain of its future in Cuba. It is cheap and easy to store at room temperature, but we don't belong to MINAG so they have not promoted it. Farmers can obtain it by telephone ordering and collection or by post, but the telephone service is limited and post unreliable.'

The vanguard transfer of eco-technology

The first step away from industrialization – that of substituting agrochemical inputs with biological ones – had been taken in the early 1990s through the expansion of Cuba's existing biological input service, the network of over 220 biological pest control production centres (CREEs) and more recent bio-fertilizer production centres, including provincial processing centres for organic matter and vermiculture. The huge costs savings involved in using these products was soon evident in terms of replacing imported inputs. Specific features of the CREEs stand out: they were diverse in nature, output and location; they received support from an upward chain in the form of ingredients and resources from provincial level and from national research centres, and from a downward chain through tapping into the distribution network of INISAV's on-farm technicians and local agricultural supply shops; they undertook direct feasibility trials with farmers; the simplest products proved to worked best, and simple instructions were provided with the product; the network united to exchange experiences; and they were incentivized by internal competition.

Because of their substitutive nature, success did not depend on any major attitude change, and the simplicity of products meant that the service could be run by semi-skilled technical staff and operate out of rural farms, training centres or as stand-alone units. Farmers were instructed on input usage in a pre-scriptive fashion. The most successful products were those that did not require a lot of skill to apply, and which could be stored at ambient – on-farm – temperature.

This top-down, transfer-of-technology approach provided an example of an organic service that could be run on private lines in other countries, demon-strating that certain organic inputs could be managed in the same fashion as industrialized. Nevertheless, the CREES encountered challenges during the Special Period: they lacked resources to operate effectively, and their products frequently could not reach the farmers owing to the lack of transport and even of carrying receptacles. During the Period, staff experienced a steep learning curve; whereas previously they had simply produced a number of different types of product, with the scarce resources available they had to become more aware of local demand in order to avoid overproduction that could not be sold. They did this by making predetermined contracts with farmers. Table 8.4 dis-plays the area treated with biological control methods over the decade 1988 to 1998, showing the impact of the crisis on usage, which was only gradually regaining its previous levels.

Table 8.4 *Area treated with biological control methods, 1988 to 1998*

Year	Area treated (ha)
1988	300,000
1992	980,000
1994	410,000
1996	605,000
1998	900,000

Source: Pérez and Vásquez, 2002

In addition to the standard CREEs, INISAV had developed three more sophis-ticated biological control factories and was planning nine more. An evaluation of these different scales showed that while the larger factories were more eco-nomical, they were less able to be fitted into the local production system and less flexible than the cottage-industry-scale CREE (Miedema and Trinks, 1998).

Attempts at changing the top-down research and extension approach

With the changing nature of farming structures, of resource use, and of differ-ing food needs, so agricultural research and extension also had to change, in terms of both content and methodology. Extension services were historically well developed; they included not only prescriptive advice and instruction from research and extension agents to farmers, but also the use of national media channels to disseminate farming news and techniques. In the late 1990s, television was broadcasting a weekly agricultural programme (*De Sol a Sol*), and a daily show (*Hoy Misma*) that included agricultural issues. Local radio broadcasted agricultural programmes around midday when farmers were more likely to be listening. With the increased awareness of the need to reach farmers, MINAG's Agricultural Information Centre, based in Havana, decentralized in 1998 to facilitate the flow and accessibility of information to a wider range of end-users through MINAG's provincial offices. It was renamed AGRINFOR and the move included computerization to reduce the need for scarce paper resources. The provincial information departments synthesized a range of technical information into more farmer-friendly terms. Some extension materials were tendered out to research institutes, and the development of technical recommendations became a collaborative effort involving researchers, farmers and other stakeholders.

Yet owing to the disciplinary nature of research institutions, there was no existing nationwide extension service. With the Special Period, a new Agricultural Extension System was planned, building on the networks of already existing extension efforts and supported by international funding. The plan would be to gradually phase in fee-based extension services, but only at a time when farmers could afford it.

More traditional extension approaches continued to play an important role where a technique or piece of information required mass diffusion, such as in the case of ox-handling, for the biological control of pest outbreaks, or for certain end-users such as state and ex-state farms that were not ready to become completely autonomous.

Traditional research had focused on the predetermined needs of the planned economy. According to staff in MINAG's Science and Technology Department, 'Researchers used to select those farmers who produced the highest yields and who had leadership characteristics but who were not representatives of the majority of farmers.' Research was deemed to have been successful if there was 'proof from the beneficiaries that the technology had worked, such as a pile of letters.' Towards the end of 1995, MINAG, CITMA and MES started a process of inter-institutional consolidation to revise the aims, objectives and strategies of the 19 research institutes. This led to the establishment of a model network: the National System of Agricultural Science and Technological Innovation (SINCITA). SINCITA's mission was 'To contribute to national food security, through the development of sustainable agriculture and the international competitiveness of the agricultural sector for the benefit of Cuban society' (Mato et al, 1999). These changes led to the introduction of formal tendering for research funding administered by the three ministries. Research projects were of four types: national, branch (based on topic), regional, or local (at the request of producers – largely state). Inter-institutional collaboration, multidisciplinarity and a focus on problem-solving became important criteria for acceptance of proposals.

Research had to refocus on techniques that could be rapidly commercialized and disseminated, and on regionally specific problems. In practice, whilst in the early part of the decade research had refocused to whatever was still available to work on, such as ways to recycle and reuse agro-industrial wastes and to test more radical approaches to production, by the end of the decade this approach had shifted to more sophisticated experimentation with organic inputs and integrated strategies. Some research areas were avoided even during the critical times, such as the use of human waste as a fertilizer. A soil fertility researcher justified this: 'There are lots of ways of stopping direct nutrient leaks without recourse to recycling human waste. For example, lots of biomass is still being burned, and also city waste. We can better increase efficiency of our nutrient recycling.' In some disciplines, there was an evident lack of institutional knowledge to identify appropriate solutions. For example, the mandate for those working in water management was to investigate ways to reduce petrol use, to make field work more efficient and easier, and to save on water. However, more ecological water management strategies were not known and not widely promoted; mulching, for example, was considered inappropriate for large farms. Instead, the focus for researchers was to improve irrigation efficiency and to find technological solutions to transport water to where it was needed. In other disciplines, more substantial shifts were made. The Research Institute for Agricultural Mechanization (IIMA), for example, had started working on alternative tillage since 1982, and with the post-crisis decline in availability of fuel and spare parts, it refocused towards animal traction implements, the promotion of traditional knowledge and integrated management, and the conservation of soil, flora, fauna and organic matter. The most costly farm activity in terms of petrol consumption was soil preparation, so to reduce topsoil inversion IIMA identified an old technique of horizontal cutting using a multi-plough (*multi-arado*). This technique had a 40–50 per cent lower energy requirement, and the horizontal cut meant that the land could be used sooner after ploughing (7–15 days compared to the previous 45–90 days). Other benefits of this technique were less soil compaction and lower incidence of weeds.

There was a general trend towards the decentralization of research units combined with a merging at provincial level of previously separate local delegations. So, for example, whereas previously each research institute in Havana had its own regional and/or provincial outposts, these outposts were now combining and taking on more responsibility. Researchers diffused results through the state Agricultural Enterprises. This required greater awareness on the part of researchers about the end users, but their lack of knowledge of, or access to, more appropriate methodologies such as participatory techniques was a constraint to its implementation. One researcher explained that: 'Scientists have the know-how but we don't know how to transfer it.' Some exceptions emerged, and a clear example of how the lack of resources instigated change and positive learning is provided by experiences of researchers in the plant breeding sector, and in particular how the method of participatory plant breeding emerged as an appropriate alternative in the existing centralized seed system. This experience is described in Box 8.2.

Some features of the pre-existing research system were retained. These included the Science and Technology Forums, whereby promising research and innovations at municipal level were selected to compete for prizes and for taking forward to the provincial-level Forum, with the goal of winning a place at the national-level Forum and possibly from there to be disseminated throughout the country.

In practice, and despite the plans and initiatives for new models, the traditional research approach was still widely used, whereby a technology was developed in isolation for 2–4 years before being applied on farm or evaluated by farmers. Where possible, research institutes attempted to work with reference farms, usually the most high-performing state or UBPC farms. Attempts to promote participation and other new methodologies were slow to be adopted and raised their own challenges. These challenges included the inter-generational conflicts as older staff resisted or were unable to adapt to new/foreign ideas being brought in by younger staff; the limited understanding of the purpose of facilitatory and farmer-led approaches and of social science in general; the mistrust by farmers of government extensionists; and the limitations that a planned production system placed on experiment and change. One more enlightened researcher explained the slow progress: 'We know this approach is not participatory enough, but it's better than the old system and takes time to change. We realize that mass diffusion is impossible, but we can't change the mentality overnight.' The concept of being a facilitator proved awkward; in Cuba a balance had to be struck between taking a back seat to stakeholder participation including that of local government, and ensuring sufficient personal recognition for success in order to justify continuation of the project and/or methodological approach. One Cuban NGO that applied the facilitatory approach found that it was being challenged for not delivering, as other stakeholders in the project were receiving recognition for their achievements. A researcher with several years' experience of pioneering more interactive extension development described the role of the 'new facilitator' of Cuba thus: 'The main input necessary is knowledge, and to integrate this with indigenous knowledge. Therefore the extensionist is a knowledge promoter. Graduates are not necessary because personality and local acceptability are more important

Box 8.2 *Pioneering localized approaches to plant breeding and seed management*

The plant breeding and seed management sector was one of the slowest to adapt to the conditions of the Special Period. The industrialized farming system with its low range of biodiversity had encouraged a vulnerable crop-genetic framework, and the heterogeneous environment that previously had been regulated through the high use of inputs was now no longer able to support the same crop varieties. The national seed system urgently needed to expand its range, but lacked the financial resources even to maintain its network of large-scale seed multipliers. Between 1989 and the mid-1990s, production capacity for seeds of several staple crops, such as maize and bean, fell by half. The solution was not as simple as one of technology substitution; the centralized breeding and seed supply system was challenged to address the real issues of enabling access to varieties that were adapted to local, low-input conditions, and the state and ex-state farms were not accustomed to selecting and producing their own seed and thus could not easily take matters into their own hands.

In the food-scarce period of the early 1990s, a young plant breeder in Havana Province was handed a challenging mandate: to rapidly increase pumpkin production. Between the late 1980s and 1993, yields of pumpkin had fallen from an average of 2–3t/ha to 0.2–0.4t/ha, and pumpkins had almost completely disappeared from the market. To increase yields required the plant breeder to compare foreign varieties with local landraces, but his research institute did not have sufficient resources for him to carry out these trials on the research station. However, on visiting some *campesino* farms that were still producing pumpkin, he realized that these farmers had the resources to support field trials on their lands. It soon also became clear that these farmers had more knowledge of crop selection criteria than the researchers, and so a participatory process was developed to involve these farmers in the research. The research team found advantages to this approach: they were able to harness farmers' capacity to identify varieties resistant to drought and pest and disease attack, and they also learned of other plant characteristics important to the local farming system of which they had previously been unaware. Results of these trials caused the research team to rethink not only their breeding strategy but also their underlying paradigm. Using pumpkin landraces selected by farmers, they were able to obtain the same high yields of 6–10t/ha under organic production as those obtained under high-input conditions, but with a saving of approximately 75 per cent of energy expenditure (based on the reduced need for irrigation, fertilizer and pesticide application). The plant breeders realized that these pumpkin landraces had been selected over generations from a highly diverse genetic pool that allowed for free cross-pollination on a dynamic basis; this contrasted with the prevailing approach of breeding and maintaining varieties in isolation. This experience led the plant breeders to develop the first programme in the country on participatory plant breeding, based on the hypothesis that to achieve high and stable yields necessitated a high range of genetic diversity within a crop. *Campesino* farmers in more isolated regions of the country, who had not been incorporated into the state seed system, were still maintaining and conserving high levels of genetic variability and this could be used to resuscitate agricultural biodiversity more widely.

Source: Ríos Labrada and Wright, 1999; Wright, 2005

characteristics. Young extensionists are particularly successful, because the farmers feel paternalistic and the relationship works.'

Learning about more appropriate farming systems

Changes in agricultural education

Formal agricultural education remained a priority and to some extent was strengthened over the decade. Schools in the Countryside (*Escuelas del Campo*) had existed since the 1970s and encouraged an understanding of agriculture amongst the youth (Rosset and Benjamin, 1994). From 1995, basic agriculture was an option on the syllabus of most primary school courses in Cuba. Additionally, ANAP started a programme to teach children within CPA and CCS cooperatives whilst they worked in the fields for half their time. At the start of the Special Period, agricultural polytechnics were instigated in every munic- ipality, each affiliated to a local Mixed Crops Enterprise. Organic agriculture was included as part of the syllabus, and this necessitated a retraining of teach- ers in this subject (Crespo and Alvarez, 1999). Every university contained a department of agriculture, and of the 600,000 college graduates in Cuba, 27 per cent held agricultural degrees (Lane, 1999). The Agricultural University of Havana (UNAH) in particular had over 10,000 part- and full-time diploma, degree and higher-degree agricultural students at any one time. This enabled continued widespread technical support for the agricultural sector, with 24,000 qualified agronomists nationwide in 1996 (Ramirez, 1997, quoted by Lane, 1999). For some institutional staff, formal retraining was organized; for exam- ple, plant health specialists underwent training on biological pest controls, and some extension staff were trained in participatory approaches. For most, how- ever, the feeling was that the greatest training needs lay with farmers, and also, as one ministry official explained, 'It would be difficult to retrain the researchers because time would be required to change mentalities.'

Recognition of *campesino* production

Throughout the Special Period, the *campesino* sector had proven its resilience through recourse to traditional practices and sufficient autonomy to innovate (as well as its stockpiling of inputs). Institutions recognized the significance of *campesino* production, and that *campesino* farmers had the capacity not only to innovate but also to evaluate and accept or reject technologies. Some found it hard to envisage that smallholder agriculture could form the basis for a self- reliant national food system. As one extensionist reflected, 'Even though I have heard that Europe is feeding itself in this manner, it can't happen here.' Other researchers and extensionists encountered challenges in working with the more autonomous *campesino* sector, especially in terms of attitude change and knowledge levels. One researcher explained: 'Farmers prefer not to use rotations and green manures because they have conventional thinking, so we cannot be revolutionary in our recommendations because it will not work in practice – farmers won't accept it.'Another commented: 'Farmers think that weeds can grow from applying manure. They should know that the manure's own process of fermentation will kill them.'

Campesino farmers were more actively encouraged to participate in state production plans and to step up production of specific staple crops such as rice, roots and tubers. *Campesino* strategies, such as rotations, intercropping, diversification and small-scale processing, were disseminated to UBPC farms. Researchers started to work with *campesino* farmers to test and adapt new agricultural technologies, and successful innovations were recognized and promoted. Such change was not found everywhere; historical support linkages continued to favour state or ex-state farms.

Of equal importance, extensionists and researchers had started to learn from farmers and their practices. This learning included that traditional practices were vitally important as the backbone for a resilient agriculture, that farmers had the capacity to innovate, that changes in husbandry practices might necessitate a change in mentality of the farmer, and that farmer autonomy was important to enable spontaneous adaptation. One social geographer, for example, observed through her own field research that 'Farmers are practising new strategies, neither stemming from the research institutes nor from traditional knowledge, but new innovations such as intercropping with plantain, or planting early varieties because of the demand from the farmers' markets.' Similarly, Guillot Silva et al (2000) documented the emergence and existence of 'spontaneous' organic agriculture, whereby farmers were building up complex, integrated and profitable agro-ecosystems without external support.

Institutional perspectives on the Special Period

More so than farmers, institutional professionals saw the Special Period as having several positive aspects and outcomes. Communications had improved both amongst institutions and with farmers. The number of international projects and training opportunities had increased. Work had become more interesting, as one soils researcher commented, looking on the bright side: 'The ensuing changes changed the monotony. It was challenging to have to invent and find alternatives, to look at more practical things and be less theoretical and abstract.'

Some felt that yields were good or had even increased, while others saw yields as being lower than previously, and these different perspectives were qualified by input levels, management style, land area and natural resource capacity. One banana agronomist explained: 'People blame the lack of petrol, but in the late 1990s yields are higher than in the 1980s. We used to fumigate twice monthly against Sigatoka – by aeroplane – but this stopped in 1992. We have 1 million tons of production, the same as in 1992, but we are more efficient now.' The degree of success of organic production was also felt to depend on the type of farm and its previous operations, as one plant breeder who had taken a course in agro-ecology commented: 'It depends on what you planted previously, such as a legume rotation, and the two best ways to achieve good maize yields in Cuba under current conditions are with good seeds and correct timing – planting after April. I recently met a group of farmers who were getting very high yields through spraying a preparation of *cachasa*, manure and water, and I would also recommend minimum tillage and 12–30t/ha manure.' Some noted the positive influence of the new markets, such as one researcher who observed that 'Farmers are practising new strategies, neither stemming from the research institutes nor from traditional knowledge, but new innovations such as intercropping with plantain, or planting early varieties because of the demand from the farmers' markets.'

The belated development of certified organic agriculture

In late 1999, MINAG made an internal announcement of its intention to produce organically for export, according to one ministry official, 'An order has come from the top to give the Ministry of Agriculture the green light to pursue organic production for export,' and by 2000 MINAG was attempting to institutionalize certified organic agriculture. That this had not occurred earlier in the decade is attributed to several factors. Organic exports could not be considered until national food deficits had been filled. Perhaps more realistically, for much of the Special Period, the agricultural sector had been simply too busy struggling to feed the population to turn to export activities. Another reason for not having developed organic exports was the insufficient organizational capacity and know-how to meet export quality standards, and the insufficient funds to build such capacity. In addition, higher-level personnel had been unconvinced of the benefits of organic agriculture.

Organic export production was included in long-term production plans for certain crops such as sugar and banana. While responsibility for developing organic agriculture lay with the Organic Agriculture Group (GAO) of the Cuban Association of Agricultural and Forestry Technicians (ACTAF), draft national organic standards were produced by MINAG's Department of Quality Control. These standards defined both organic and sustainable production methods, the former depicted as being environmentally friendly and bound by legislation, and the latter focusing on meeting societal needs and with the regulated use of agrochemicals.

As of 2000, Cuba had no permanent, commercial-scale organic farms, although throughout the decade foreign buyers were intermittently certifying the production of specific crops (largely sugar and fruits) in scattered locations, for export through agreements with MINAG and the Ministry of Foreign Trade. This export-driven production was isolated from the rest of the farm upon which it was based, and was only sustained by the interest of the buyer. Certified organic crops being developed were sugar, citrus, banana, coffee, honey, urban horticulture, medicinal plants, coconut, pineapple, cocoa and mango. Box 8.3 describes the experiences with the development of organic citrus.

The learning experiences with organic citrus production exemplified how the so-called 'organic-by-default' production of Cuban agriculture, even in the absence of chemicals, was vastly different from the type of production required to meet organic standards. In parallel with this, MINAG was considering how to reach the domestic tourist market through an own-brand, softer labelling system, which would prohibit the use of agrochemicals but would not require international accreditation. This brand would still prohibit the use of agrochemicals and attract a premium price. Products could be channelled through the normal tourist supply routes of Frutas Selectas and Citricus Caribe. A few pilot projects were already supplying 'productos sanos', or healthy produce, to national tourist markets, and several Cuban institutes were interested in becoming involved in these projects because of the potential dollar revenue to be earned.

Box 8.3 *Experiences with certified organic citrus production*

Citrus was the commodity sector where most experience had accrued on certified organic production. This was partly because of the longstanding commercial expertise of the Cuban Citrus Corporation; as it was a major export crop, citrus production and marketing was handled by a specialized commodity corporation rather than through MINAG's Agricultural Enterprises. Citrus being a minor staple of the Cuban diet, there were lower domestic self-sufficiency goals to fulfil before export than for major staples such as banana. The Citrus Corporation itself had a longer vertical structure than other state Enterprises, topped by a greater degree of institutional autonomy. All the required administrative and commercial functions were contained within the Corporation itself rather than located externally in other state organizations. This made it easier to build internal consensus, and to incentivize and cooperate to achieve agreed goals without the challenge of inter-institutional bureaucracy. This, and the Corporation's historically accrued financial and managerial skills, and resources and export experience, gave it a relative advantage for developing an organic export line.

In 2000, the Corporation formed a multidisciplinary, organic group to undertake research and development on the whole production chain from field to market. The formation of this group did not require much change in research orientation, because researchers there had for a long time been working on organic issues – such as bio-pesticides – albeit in a disciplinary fashion. The Special Period acted as a catalyst to pull these disciplines together. With an increase in international demand, some of this team were supported to undergo specific training on organic agriculture, and support and advice was received on an ongoing basis from abroad. The project liaised with other national institutes on specific system components; with the Institute for Pastures and Forages on under-sowing legumes, with the Institute of Mechanization on appropriate equipment such as compost spreaders (which the project was helping to evaluate), with the Institute of Ecology and Systematics on local biodiversity characterisation, and with the Soils Institute for a supply of compost.

The Corporation's approach to organic citrus production was aimed at developing a large-scale and intensive system with high production goals. At 2000, several citrus cropping areas on UBPCs in Havana, Cienfuegos and Ciego de Avila were starting with the conversion process. These UBPCs had been selected on the basis of their extent of diversification, their proximity to industrial centres and their previously low levels of use of agrochemicals. Although the crop itself was cultivated as a monoculture, a whole-farm approach was adopted as far as possible, to create a semi-closed, self-reliant farm system. This meant, for example, that manure was sourced from on-farm livestock, which overcame the increasing difficulty in accessing manure from other sources. Compost was made from the residues of the citrus processing mixed with livestock wastes, and bio-fertilizers and zeolitic organic fertilizers also added. Application of these techniques served not only to eliminate the use of agrochemicals, but also had an appreciable affect on yields and fruit quality, on economic returns, and on the conservation of natural resources. Yet the forecasted economic returns did not internalize the high start-up costs, which included the time and effort required to learn and develop the new techniques and technologies.

Citrus farmers were trained on new technological issues such as vermiculture, compost, legumes, and the use of Rhizobium and Azobacter. The extension approach was described as interactive, through 'a discussion of what is being transferred'. Farmers were apparently receptive to an organic approach owing to its similarities with traditional family farming, and they were noting improvements in plant health. However, given the high start-up costs, farmers were concerned about profitability. Production costs were also high in the first year of conversion, owing to the plantations having received very little care previously. It was anticipated that labour demand would ease off after conversion, and that the high quality and care of production would mean fewer postharvest losses. Nevertheless, production was unlikely to be profitable in the conversion period, and the Citrus Corporation was bearing the losses. The issue of certification had not yet been addressed.

Overall, the project was turning out to be a challenge, as those involved learned that certified organic production was not synonymous with low- or no-input, but that it required building complexity and ensuring quality also of organic inputs and their sources.

Source: Wright, 2005

There was no certified or non-certified organic production for domestic markets. Even if the purchasing power existed, the state food collection and distribution system challenged product differentiation. However, another possibility being considered by MINAG at the end of the decade was for the agrochemical-free production of selected crops for the Cuban people, at least in small volumes. These products would not require standards and could be distributed through normal channels with minimal price premiums.[2]

The concurrent growth of the biotechnology industry

At the outset of the Special Period, the state made an investment of $100 million into biotechnology research, for both agriculture and medicine. Biotechnology was seen as a major potential generator of foreign revenue. This industry, based at the Institute of Genetic Engineering and Biotechnology (with 1132 staff), included the development of biological pesticides and fertilizers, but its biggest strength was in genomics. Although there were no nationally produced transgenic crops in commercial production or on the market in 2000, the medium-term plan was to develop such crops for domestic, non-tourist consumption, to be distributed through the ration system and for use in processed products. Food destined for the tourist sector would be kept GM free, and GM development was not applied in the tobacco industry for fear of harming international sales. Figure 8.1 sets out this plan.

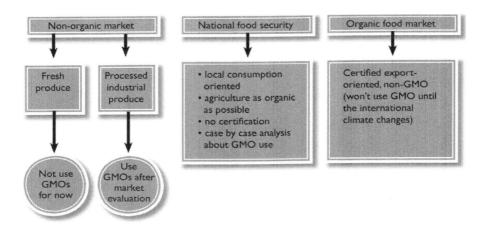

Figure 8.1 *Plan of the Biotechnology Research Centre for the use of GM crops in Cuba*

Source: Wright, 2005

Owing to food safety concerns, transgenic foodstuffs such as soya, although cheap, were not yet being imported, and transgenic material imported for research purposes was subject to a laborious registration process. Research was currently under way to evaluate the implications of transgenic crops for human health and the environment. Given the difficulties in accessing the internet and international journals, Cuban scientists were relatively cut off from international debate on transgenes, and any new technological advance was of immediate interest. Information on the potential risks of GM technology was, however, being introduced by the Cuban organic movement (e.g. Fernández et al, 1999). For the industry, it was not so much a question of 'if' genetic engineering would go ahead but 'under what circumstances'.

The emerging components of a new type of agricultural system

In the early 1990s, desperation had allowed for a 'try anything' approach which embraced support for the recycling and re-use of agro-industrial wastes, as well as the testing out of alternative agricultural concepts such as permaculture. By the late 1990s, a more sophisticated approach prevailed that was focused around a broadly interpreted concept of integration. Within this, some components of the agricultural support system still adhered to the industrialized perspective, whilst other components embraced an alternative, localized, ecological perspective, and these co-existed in contradictory fashion. Some remnants of the industrialized system remained but were requiring change, such as the top-down extension model and the centralized seed system. Other components of the industrialized system were kept for their continued applicability, such as the supply of agrochemicals. Similarly, some more ecological components had been adopted prior to the Special Period, such as research on biological pest controls, others were present by default, such as the new on-farm research approaches, and others still had been deliberately chosen as part of a new system (although may have also first arrived through lack of choice), such as urban agriculture. Figure 8.2 schematizes this dynamic of coexisting, contradictory components.

Viewing the emerging agricultural support system in this manner enables the identification of components that are obsolete or even hindering development and that could be discarded, and also components that are beneficial and could be further institutionalized and scaled up.

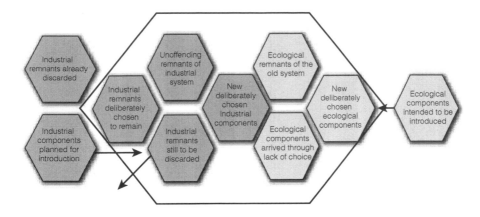

Figure 8.2 *Coexisting industrial and ecological components of the emerging agricultural support system*

Source: Wright, 2005

Notes

1   As with farm cooperatives, it was obligatory in institutes that management positions were rotated, which in practice meant encountering relatively young staff in high positions and ex-management subsequently serving in the lower ranks. As one aging ex-director commented, 'I think its an error to allow older people to stay in positions of power. Even Fidel has a substitute – everyone has it, to continue the Revolution.'

2   In the province of Cienfuegos, *organoponicos* were producing quantities of vegetables under a production scheme called 'functionally organic'. However, this scheme used certain inputs that would not be permitted within international organic standards (Taboulchanas, 2001).

References

Castro, A., Marrero, L. and Arias, A. (2000) 'Potencial productivo y composición nutritiva del grano de cuatro variedades Cubanas de sorgo cultivadas en condiciones de bajos insumos', *Proceedings, IV Symposium on Sustainable Agriculture of the XII Scientific Seminar of INCA*, 14–17 November, INCA, Havana

CITMA   (2000)   Scientific   potencial.   www.cubagob.cu/ingles/des_soc/sitiocitma/potential.htm accessed in May 2000

Crespo, A. and Alvarez, F. (1999) Presentation of the Director and Vice Director. IP School Tranqilino Sandalia de Noda, Food First Delegation tour, 21–28 February 1999

Díaz, S. (2000) 'La dimensión ambiental en la proyección estratégica de la ciencia y la innovación tecnológica agropecuaria en Cuba', Conferencias Magistrales, *Proceedings, IV Symposium on Sustainable Agriculture of the XII Scientific Seminar of INCA*, 14–17 November, INCA, Havana

FAO (2003) *Fertilizer Use by Crop in Cuba*, Rome, Land and Water Department Division, FAO

Fernández, L., González, T. and Fundora, Z. (1999) 'La Biotecnología y sus riesgos', *Agricultura Orgánica*, vol 5, no 3 (December), pp136–137

Garcia, M. (1998) *Contribución al studio y utilización de los abonos verdes en cultivos económicos desarrollados sobre un suelo ferralítico rojo de La Habana.* Unpublished PhD dissertation, Havana, UNAH

Guillot Silva, J., Caballero Fournier, O. and Acuna Serrano, B. (2000) 'Una finca campesina ejemplo de integración agroecológia espontánea', paper prepared for *IV Encuentro de Agricultura Organica* (postponed), ACTAF, Havana

Lane, C. S. (1999) *Agricultural Education and Extension in Cuba During the 'Special Period'*, Dep. Educational Policy and Administration, University of Minnesota

López, V., Mastrapa, E., Nunez, R., Martínez, R. and Aranda, S. (2000) Diagnóstico sobre la implementación del manejo integrado del tetúan del boniato (*Cyclas formicarius var. Elegantulus*) en la Provincia del Holguín', *Proceedings, IV Symposium on Sustainable Agriculture of the XII Scientific Seminar of INCA*, 14–17 November, INCA, Havana

Mato, M. A., Maestrey, A., Muniz, M., Alvarez, A. and Fernández, M. A. (1999) *La Consolidación del Sistema Nacional de Ciencia e Innovación Tecnológica Agraria (SINCITA) del Ministerio de la Agricultura (MINAG) de Cuba: Experiencias, lecciones e impactos de un proceso de cambio institucional*, La Habana, Ministerio de la Agricultura

Miedema, J. and Trinks, M. (1998) *Cuba and Ecological Agriculture. An Analysis of the Conversion of the Cuban Agriculture towards a more Sustainable Agriculture.* Unpublished MSc thesis, Wageningen Agricultural University

MINAG (1998) *Instructivo técnico sobre el cultivo del boniato/ malanga/ calabaza/ ñame/ yuca*, Habana, SEDAGRI, MINAG

Monzote, M., Muñoz, E. and Funes, Monzotc F. (2002) 'The integration of crops and live-stock' in Funes, F., García, L., Bourque, M., Pérez, N. and Rosset, P. (eds) (2002) *Sustainable Agriculture and Resistance: Transforming Food Production in Cuba*, Oakland CA, Food First Books, pp190–211

Pérez, N. and Vázquez, L. L. (2002) 'Ecological pest management', in Funes, F., García, L., Bourque, M., Pérez, N. and Rosset, P. (eds) *Sustainable Agriculture and Resistance: Transforming Food Production in Cuba*, Oakland CA, Food First Books, pp109–143

Ríos Labrada, H. and Wright, J. (1999) 'Early attempts at stimulating seed flows in Cuba', *Magazine of the Information Centre for Low External Input and Sustainable Agriculture (ILEIA)*, vol 15, no 3/4 (December), pp38–39

Rosset, P. and Benjamin, M. (1994) *The Greening of the Revolution. Cuba's Experiment with Organic Farming*, Ocean Press, Melbourne.

Sinclair, M. and Thompson, M. (2001) *Cuba Going Against the Grain: Agricultural Crisis and Transformation*, Oxfam America. http://www.oxfamamerica.org/newsandpublications/publications/research_reports/art1164.html accessed May 2002

Taboulchanas, K. H. (2001) 'Oportunidades para la certificacion organica', poster presented at the *IV conference on Organic Agriculture* (postponed), ACTAF, Havana

Treto, E., Garcia, M., Martínez Viera, R. and Manuel Febles, J. (2002) 'Advances in organic soil management', in Funes, F., García, L., Bourque, M., Pérez, N. and Rosset, P. (eds) *Sustainable Agriculture and Resistance: Transforming Food Production in Cuba*, Oakland CA, Food First Books, pp164–189

9

Perspectives on the Mainstreaming of Local Organic Food Systems

Driving forces behind current levels of agricultural sustainability

Several foreign reporters have commented on Cuba's pro-organic policy. 'Organic agriculture has been adopted as the official government strategy for all new agriculture in Cuba, after its highly successful introduction just seven years ago' (*The Pesticides Trust*, 1998); 'Cuba is perhaps the best example of large-scale government support to organic agriculture ... It is an integral part of agricultural policy' (Scialabba, 2000). In reality there was, up to 2000, no legislation on organic agriculture, certified or otherwise. Perhaps the most firm, high-level endorsement was in a speech given by Castro at the 1996 World Food Summit, where he stated that 'Enhancing food security demands extending sustainable agricultural techniques so that the various economic units operate as agro-ecological farms.' Therefore, although certain elements of an organic system were present in the country, such as the CREEs, the grassroots organic movement (ACAO), organic training courses, urban *organoponicos* and so on, there was no policy to gel these elements together nor to prioritize these over other strategies. In this sense, any semblance of a widespread organic system was in place, not through cohesive policy but through a shared lack of other options.

The driving forces for the increase in organic production approaches in Cuba over the 1990s were the same as those that united and directed agriculture nationwide: the lack of fuel and agrochemicals, and the need for self-sufficiency. The crisis of 1989, together with imposed sanctions and political isolation, instilled the need for national independency from externalities, including of inputs. At regional and local levels too, independence from external inputs played a motivating role; organic products were produced in-country and were cheaper, and on-farm input substitutes were more readily available, more economical and more secure.

The Cuban experience indicates that a scarcity of agrochemicals and fuel does not necessarily lead directly to a widespread, localized organic production strategy. Further proactive mechanisms would be required, not least concerted policy backing, and other factors were at play that prevented or constrained these.

The agricultural sector in transition

In tracking the development of organic farming in Cuba, Funes (2002) identifies that 'the principal techniques receiving widespread application have only been of the "input substitution" or "horizontal conversion" varieties'. He refers to the period of the 1990s as the 'first phase', the basis for further widespread consolidation of organic agriculture. In practice, different individuals, groups and institutions were occupying different positions in terms of farming approach; some were still operating along industrialized lines, some were substituting agrochemical inputs for biological inputs as suggested by Funes and others, while a minority had transformed further to give up a reliance on any type of input and instead focus on a balanced interaction with nature. At the extremes was a tendency for ministerial institutions to operate in a more industrialized fashion, compared to organically oriented projects organized by pioneering farmers, researchers, extension groups or NGOs. Figure 9.1 depicts this transition from more industrial to more organic agriculture for the various groups within the agricultural sector.

Not everyone started from the same position: some pioneers, such as the organic movement (ACAO/GAO) and the IPM researchers, were already operating along organic lines early in the 1990s, whereas others, such as the state farms that converted into UBPCs, or the farm mechanization network that

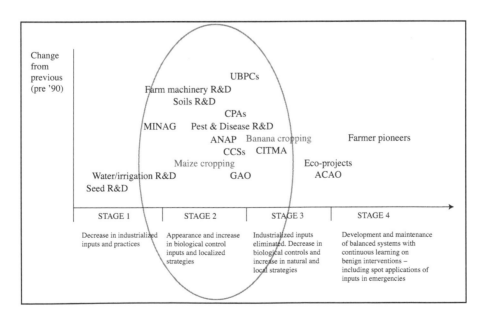

Figure 9.1 *The organic transition for different disciplines, institutions, farmers and cropping systems*

Source: Wright, 2005

embraced animal traction, had made huge strides since 1994 to shift to organic techniques. There were also differences in the regimes for different crops; non-prioritized crops such as maize had remained low-input, whereas crops such as banana had modified from high chemical input to a relatively more organically oriented production regime. Thus, although the agricultural sector as a whole was in the phase of input substitution, many individuals and groups had nevertheless undergone huge transformations from their starting points at the beginning of the decade. Others had not. In the transition process, workers in the sector noted two points where a shift in attitude was required: the first shift came in the letting go of agrochemicals in favour of organic inputs, and the second was required in order to accept that even organic substitutes were unnecessary and to make way for more benign management approaches.

Interpretations of organic agriculture and its potential for mainstreaming

The Cuban interpretation of organic agriculture

Although specific individuals – farmers and scientists – had been working with organic principles since the 1970s and 1980s, organic agriculture emerged in Cuba as part of a broader response to the food security crisis, symbolized by the formation of the Cuban Association of Organic Agriculture (ACAO) in 1993. The emergence and achievements of ACAO are described in Box 9.1.

Because of the context within which the organic concept emerged, the concern of the movement was not on the avoidance of agrochemicals or on market returns, but rather on improvements to the production system based on ecological principles, to adjust the prevailing industrialized approach for increased yields. Key interests were the integration of the previously specialized and separate crop and livestock farms, increases in energy, fuel and land use efficiency, and reversal of the degradation of the natural resource base. For example, Monzote et al (2002, p207) described crop-livestock integration thus: 'This concept has implications which go beyond the technological-productive sphere, directly or indirectly influencing the economic, social and cultural conditions of farming families, by reinforcing their ability to sustain themselves.' Because of its basis in research, the Cuban interpretation of organic agriculture was both cutting-edge and pragmatic, and the concept of equilibrium was routinely used. This interpretation conformed with the Latin American agro-ecological school of thought rather than the European certified organic model.[1] In contrast, the term 'organic agriculture' held certain negative connotations associated largely with the expense of certification and the higher price of certified organic products. As one rural sector worker put it, 'Organic agriculture is based on

---

Box 9.1 *The Cuban organic movement (ACAO) and its achievements up to 1999*

Unlike grassroots organic movements in other countries, ACAO was not farmer-led, but created as a response to the crisis by organically minded individuals involved in agricultural research and training. Through the decade, the movement grew to approximately 800 members nationwide, including farmers, and was instrumental in many pioneering initiatives. It had no juridical status and was dependent on donor funding and on the voluntary input of its members. In 1996, ACAO received the Sustainable Agriculture and Rural Development (SARD) Prize for its work and, in 1999, the prestigious Right Livelihood Award, which it accepted on behalf of the nation. Its achievements were impressive under the conditions of the Special Period, and included:

- awareness raising and information diffusion through workshops, field days and other events
- operation of mobile organic libraries around the country
- tri-annual publication of a national magazine *Agricultura Organica*
- initiation of a pioneering organic research and demonstration project 'Agro-ecological Lighthouses' to evaluate the efficacy of organic approaches
- organization of three national organic conferences
- supporting the establishment of higher education programmes on agro-ecology
- organization for the training of Cuban professionals by foreign organic certification bodies
- development of strong international linkages including with the international organic movement
- establishment of provincial support groups for organic agriculture
- initiating the development of national organic standards
- promotion of farmer-to-farmer, agro-ecological extension methodology with ANAP.

Source: Wright, 2005

---

bio-inputs and is expensive because of certification, whereas agro-ecology may include the use of agrochemicals if absolutely necessary.' Another explained: 'There is no alternative to sustainable agriculture. Both organic and Green Revolution agriculture are like agribusiness.' Rather, some professionals felt that Cuba exceeded the expectations of organic farming in the west: 'In fact Cuba talks about ecological agriculture which is one step beyond organic agriculture – it has to show sufficient yields to solve the food crisis. The conventional model can't solve the problem, but the agro-ecological model can, through a slow but steady process of increasing yields and quality.'

Orellana et al (1999) carried out an interesting survey in Cuba on organic perspectives, in which the researchers posed three questions to a group of 72 people, largely comprising scientists but also farmers and academic interest groups. Their first question asked whether organic agriculture was based solely on the use of organic inputs. This was affirmed by 60 per cent of respondents, who explained that if the inputs were not organic then it was a fraud. The large minority who disagreed felt that organic agriculture could include the use of chemicals when justified, for economic or ethical reasons. The two underlying factors that, for them, defined organic agriculture were soil fertility and biological productivity. The second question asked whether organic agriculture could feed the needs of a growing population: 54 per cent of respondents agreed that it could. The third question, on whether urban agriculture was grounded in an ecological basis, was affirmed by 78 per cent of respondents. From this the authors concluded that organic agriculture was 'a management system in harmony and dynamism with the agro-ecosystem, which guarantees an integrated protection of foodstuffs with high biological quality'.

Emerging evidence on the benefits of organic production approaches

Farmers' experiential evidence on the performance of organic agriculture

Farmers were clearly aware of the benefits as well as the negative impacts of the decrease in agrochemical use: benefits to their own health as well as that of the soil, to livestock and the natural environment. Since agrochemical use had been limited, incidences of poisonings, allergies and skin diseases had disappeared. According to one CPA farmer, 'Those who work in fumigation can only work for five years maximum in that job and then have to change, because of illness – vomiting, nausea and lack of energy.' Knowledge levels had dramatically increased, and cooperative workers and farmers were more engaged with their work. Immediately after the crisis, during the early to mid-1990s, farmers had noticed a rise in pest and disease numbers. As one farmer described, 'We had whitefly and *Thrips palmi* because there were fewer pesticides and those which there were arrived late.' Some had understood that the ecological balance would restore itself, and noted that it had taken five years, from 1993 to 1997, for the first visible improvements in wildlife and natural systems to occur. Economic status had improved, especially for UBPC farms. Labour requirements had increased for some farmers and decreased for others. Product quality was felt to have decreased, quality being judged by the size of the produce, which had decreased without agrochemicals. Figure 9.2 displays the key impacts of the lack of agrochemicals on livelihood aspects of farm cooperatives, as perceived by farmers.

Any resulting decreases in soil fertility were put down to poor management coupled with a low use of organic fertilizers, which contributed to erosion. Particularly in the east of the country, and with the absence of chemical fertilizers to maintain a mantle of fertility, the extent of soil degradation due to

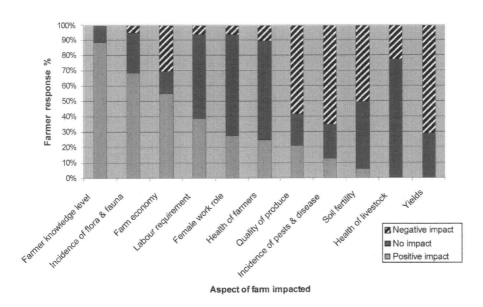

Figure 9.2 *Farmers' perceptions of the impact of the decrease in agrochemical availability on livelihood aspects of their cooperatives during the Special Period*
Source: Wright, 2005

earlier industrialized practices had become more apparent. Yet soil fertility showed an increase wherever proactive organic measures were being taken.

From the farmers' perspective, many, if not all, of their crop and livestock production lines could perform well under organic management systems, on condition that there were no external disruptions to the system such as heavy pest attack, and that there was ready access to organic inputs. In this case, organic production was technically feasible and could deliver good yields. Comments from CPA farms included 'The reality of the 1990s has demonstrated that it is possible to increase production without agrochemicals' and 'If it is managed to protect from erosion and to have organic fertilizer and good seeds, then yields will not be affected.'

Scientific evidence on the performance of organic agriculture
Economic evaluations of organic production were indicating substantial cost benefits. Early evaluation of the CREEs showed the production costs of the cottage-industry manufacturing of biological pest and disease controls to be less than 1 per cent that of importing agrochemicals to do the same job, and the money spent stayed in Cuba and helped build up the local economy (Maura, 1994). The use of green manures was found to save $31 to $75 per hectare (623 to 1503 Cuban pesos), depending on the crops and species used. Both higher crop yields and reduced outgoings for chemical fertilizers contributed to this

saving, which did not include the longer-term economic benefits of improved soil properties (Garcia, 1998). Intercropping reduced total energy use and provided a better rate of economic return than monocropping (Quintero, 1999). The combined use of organic approaches and appropriate crop varieties achieved yield levels comparable to industrialized approaches, yet the organic approach reduced energy expenditure and therefore increased farmers' profit margins (Ríos et al, 2001). In dairy systems, milk production cost 4.3c/litre from no-external-input systems, compared to 5.5c–6c/litre from systems using purchased feedstuffs (Monzote et al, 2002). In mountainous regions, the cost benefits of constructing drainage bunds and using green manures were calculated as $161/ha for maize production and $3985 for tobacco over three years (Instituto de Suelos, 1998).

By the end of the decade, a range of research findings had also emerged on the environmental, agronomic and social potential of organic agriculture. The majority of these research projects would not have been developed had the sector not been forced to reorient under the conditions of the Special Period. These findings included the following (unreferenced findings indicate direct communications with researchers):

- The highest number of mycorrhizal strains were found in the soils of *campesino* farms rather than on state or ex-state farms.
- The combining of two different bio-fertilizers could increase yields more than the use of one alone.
- Bio-fertilizers could provide up to 80 per cent of nitrogen needs and 100 per cent of potassium needs for a range of major crops (Martínez Viera and Hernández, 1995; Riera et al, 1998).
- Bio-pesticides did not leave significantly harmful residues in crops.
- The incorporation of fast-growing shade crops in a rotation acted as a weed suppressant (Paredes, 1999).
- The use of appropriate maize varieties under organic conditions produced yields of 4.5t/ha – equal to the state's high-input maize production plans.
- Intercropping had multiple benefits, for farm diversification, total productivity, efficiency of land use, and pest and weed control (Quintero, 1999; Casanova et al, 2002; Pérez and Vázquez, 2002).
- A new plough design – the multi-plough – avoided topsoil inversion and compaction, saved 40–50 per cent of petrol consumption, could be used with oxen, and resulted in productivity increases.
- Plants inoculated with Rhizobium were more resistant to both pests and drought.
- Substituting manure (20t/ha) or sugarcane residues (40t/ha) for the recommended dosage of mineral fertilizer increased maize yields by one-third, with the cane outperforming manure (Pozo et al, 2000).

- The mixing of slurry with zeolites (aluminosilicate minerals) reduced the risk of environmental pollution.
- Inoculating tree seedlings saved substantial costs on nursery activities through the doubling of germination rates (Garcia Ramirez, 1999).
- Some biological controls (such as *Telenomus sp.* and *E. plathyjpenae*) were more effective in controlling maize leaf-borer than chemical insecticides.
- Neem extract worked not only as a crop pest control agent, but also in the control of ectoparasites in livestock and domestic animals (Pérez and Vázquez, 2002).
- Molasses traps were successful in encouraging the lion ant (*Pheidole megacephala*), a predator of sweetpotato weevil.
- Fifty-three crop varieties were identified that were resistant to nematodes and with a potential for incorporation in rotations (Pérez and Vázquez, 2002).
- *Trichoderma* was successfully used to control *Thrips palmi*, helping bring yields back to pre-infestation levels.
- The incorporation of green manures gave yield increases of 4t/ha for potato and 3.5t/ha for squash (Garcia, 1998).
- During the organic conversion of farms, energy and labour efficiency increased over time, and total labour input decreased as the farm became established (Monzote et al, 2002).
- Integrated crop-livestock farms had higher levels of biodiversity (Monzote et al, 2002).

A longer-term study on whole farm systems had been established in 1994 by ACAO, termed the Ecological Lighthouse Project. Spread over three CPA cooperatives in Havana Province, the objective was to convert one area or *finca* of each cooperative to organic production for comparative research. An evaluation of the project after three years showed that organic techniques had increased total productivity by between 7 and 14t/ha, whilst eliminating the need for agrochemicals and increasing crop diversity four-fold. Soil organic matter had risen significantly (ACAO, 1999).

These scientific results were considerable, given that the scientists were working within a very resource-restricted environment. Although not mainstreamed, the results suggested that scaling-up of organic practices was technically feasible and held significant benefits for further sustainable improvements of food production in Cuba.

Future trajectories

When looking to the future, farmers tended to agree that they would use 'only the necessary' chemicals. Many saw organic as possible if certain conditions were satisfied: the availability of substitute inputs, fuel, economic incentives, knowledge and training, and will. Most professionals in the agricultural support sector agreed that there would not be a return to the high input levels of the past. They had learned that an integrated, environmentally aware approach resulted in similar or higher yields, and also that input use varied according to local agro-climatic conditions. Out of these perspectives, two potential trajectories could be discerned:

1   A return to agrochemicals and a less-industrialized model.
2   A modest return to agrochemicals and a 'smart', knowledge-based organic model.

Backing the former trajectory were those who felt that organic techniques could not perform in terms of yields. Backing the second were those who felt that the experiences of the Special Period, and levels of education and knowledge, would mean a long-term development of an organic system. Yet this would require the continuation of 'a deep-rooted paradigm shift, already underway, allowing agronomists and farmers to view the soil as a living subsystem of an agricultural ecosystem that operates according to the laws of nature' (Treto et al, 2002, p184).[2]

A high-ranking professional predicted a stronger organic policy and added: 'It is not just a group of crazy people any more. Cuba is further along the road than just doing it out of necessity.'

Influence of organic production on food availability

Several individual elements of an organic approach were being employed in the food system: of localizing production–consumption linkages and enabling the availability of more diverse and fresh produce. With over 80 per cent of agricultural land under permanent crops and pastures, so the remainder was forced under organic-style rotations and biological input use in order to increase the number of short-cycle food crops. However, the steady increase in food production could be attributed to these but also to a whole raft of other factors, and the organic-style of production employed for some crops, such as people's rice, was not easy to assess in terms of productivity gains owing to the informal and complex marketing and self-provision systems that operated around these.

Opportunities for scaling-up organic production

Factors affecting scaling-up

Funes (2002) identified a number of factors that favoured the development of organic agriculture in Cuba. These included the high number of qualified personnel, widespread experience in community approaches, supportive administrative and social structures, government-sponsored publicity campaigns in the interests of the people, favourable research findings, and the presence of organizations dedicated to the creation of an organic culture.

Those working in the agricultural sector – researchers, extensionists, management and policy-makers, and especially farmers – were clear about the type of support necessary to increase organic agriculture in the country, and even bring it into the mainstream. Farm assets such as land and labour were not considered to offer any serious constraints to organic production, although there were regional differences in this respect. In fact, any increase in labour demand that resulted from a change to organic production was seen as a benefit; farmers could employ members of their extended families. Certain physical assets were in poor condition, such as machinery, and technical information in the form of leaflets and booklets was lacking, yet neither were these considered to be major constraints. The improved economic status of farmers outweighed these and other considerations. In order for organic scaling-up to occur, farmers clearly identified two major production constraints: a lack of organic inputs; and a lack of knowledge on organic techniques.

Crucially, both these constraints were conditioned by the lack of fuel. Whatever the farmers' intention, the lack of fuel could bring their whole system to a virtual halt. Based on typical components and experiences of farming cooperatives at the end of the 1990s, two scenarios demonstrate the critical influence of fuel. Figure 9.3 depicts these: one where the farmer and agricultural support professionals were able to access fuel; and the other showing the reverse, where fuel shortages caused a system's breakdown in terms of the synergies and inputs required to maintain production.

Without fuel, farmers were unable to transport inputs such as manure from the dairy farms, bio-pesticides from the CREEs, or sugarcane waste from the CAI. Researchers, agronomists and other state support were similarly unable to reach the farmer with seeds, advice or other goods and services. Farmers were less able to travel to share and exchange seeds and knowledge with other farmers, especially if land holdings were relatively large. Neither could the farmer operate petrol-driven irrigation systems. The only aspects that could be maintained were exchange with close neighbours, agricultural programmes delivered through radio or television, and manure from any cattle kept on farm for self-provisioning. Seeds could be saved but ran the risk of theft if left

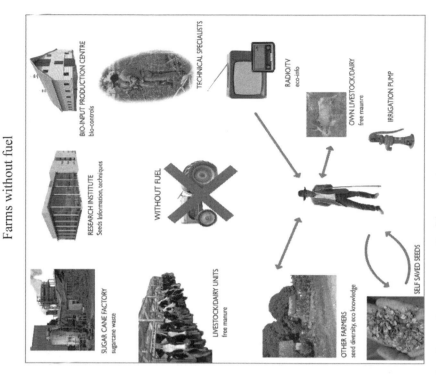

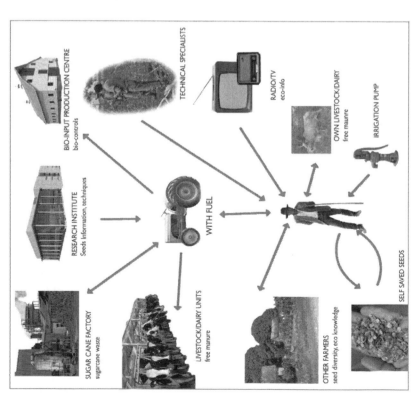

Figure 9.3 *Farms with and without fuel: maintaining or breaking the synergies that support farm cooperatives*

Source: Wright, 2005

outdoors to dry. The overall message from this was that the more self-sufficient the farming entity, in particular with regard to the complexity of re-use and recycling of farm by-products, the less affected it was by the fuel shortage.

Within this context, the opportunities and challenges to scaling-up organic agriculture fall into three clusters: those relating to knowledge; those relating to resource and technology access; and those relating to socio-political factors.

Developing ecological knowledge systems

The need to increase ecological literacy

The extent of, and capacity for, ecological innovation and experimentation was dependent on the presence of relevant knowledge. The lack of relevant knowledge and training was one of the main limitations to the increased use of organic approaches. Farmers had most commonly received training in bio-pesticides, with some also on bio-fertilizers. The thematic knowledge gaps amongst both farmers and the institutional support sector were those relating to water conservation and usage, to product quality, and to the principles underlying organic agriculture. Many farmers had not heard of the terms 'organic' (*organico*) or 'agro-ecology' (*agro-ecologia*). Farmers also identified specific training needs on the following: dietary and health requirements of draught oxen, the demonstration of appropriate soil fertility techniques, seed exchange and seed quality control, and the correct use of biological pest control products. In particular, increasing the training opportunities on organic inputs would serve to dispel farmers' concerns over the efficacy of pest control products and over the risk of disease spread through the use of organic matter and off-farm compost. Such concerns were currently limiting the usage of these inputs. Although not flagged up by farmers, training on sustainable water management would provide them with an alternative to dependency on petrol-driven irrigation equipment and might liberate those farmers who currently felt constrained by the fuel shortages in terms of productivity.

At the same time, increasing ecological literacy would serve to demystify and avoid the common misperceptions surrounding organic agriculture. Organic agriculture was directly equated with low-input agriculture, or a wealth-deficient situation. In fact, it was also referred to as 'low-income' agriculture. Agrochemicals, meanwhile, were associated with more affluent times. The logical train of thought arising from this was to avoid a production system that yielded less, especially given the state policy to maximize yields. Figure 9.4 depicts this train of thought, which was in reality proving to be inaccurate.

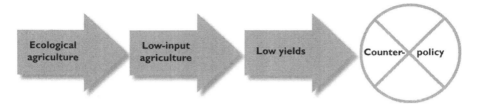

Figure 9.4 *The train of thought leading to hesitancy over mainstreaming of organic or ecological agriculture*

Source: Wright, 2005

For farmers, equating organic agriculture with a lack of agrochemicals gave rise to several speculative concerns. These concerns, or fears, were over the possible increase in pests and diseases and decrease in yields; the possible drop in product quality (quality being equated with product size); the possible increase in workloads (owing to the lack of fuel); the possibility that organic techniques were inappropriate for larger farms; and the conjecture that soils had been irreparably degraded to the point that only chemicals would suffice. Owing to their lack of ecological knowledge, farmers were not aware of its ability to enrich and regenerate the natural resource base.

Similarly, amongst researchers and extensionists, organic agriculture was associated with low yields, self-provisioning and a general poverty situation. Low-input strategies were being promoted that frequently entailed not only zero chemicals but also zero irrigation and mechanization, and were directed at marginal lands. As one extension officer explained about criteria for the selection of farmers to incorporate into organic projects, 'We will select the producers who have received fewer benefits and who have less potential, to show that if these farmers can achieve success than anyone can.' This selection affected project performance.

As well as the above concerns, a huge diversity of opinions existed on the actual performance of organic husbandry practices. These opinions, of farmers and institutional staff, were based on personal experience, on research results, and/or on secondary evidence and hearsay. For example, some in the agricultural sector felt the practice of intercropping to be detrimental as it resulted in lower yields and had a negative effect on harvest of the main crop, while others were clear that it provided a more regular income and made more efficient use of the land and of farmers' time. Similarly, some felt that the seed supplied by the state was of better quality, while others put their trust in farmers' self-saved seed for guaranteed quality and economy. This diversity highlighted the constructivist way in which agricultural decision-making was based on a personalized logic and, sometimes, on myths and misperceptions. Table 9.1 provides some examples of this diversity.

Table 9.1 *Selection of contrasting opinions on the performance of organic agriculture based on experiences in the Special Period*

Production issue	Opinions favourable to organic production	Opinions challenging organic production
Performance of appropriate crop varieties	Low-input maize varieties under organic conditions outperform state-selected varieties under industrialized conditions (Research: Ríos Labrada et al, 2000b)	Low-input varieties are not useful because as inputs drop, so do yields (Seed Enterprise staff)
Potential of farmers to self-manage seed stocks	Farmers displayed capacity to manage selection, multiplication, conservation and marketing of seeds (Research: Ríos Labrada and Wright, 1999) 'I was taught in one session how to cross my own basic maize variety with another' (Farmer) Seed saving is useful to avoid dependency on the state, to improve quality, economic efficiency, yields, germination rates and appropriateness. (Farmer) Farmers are capable of maintaining quality control (ANAP staff)	Farmers have difficulties in selecting varieties (Plant breeder, MINAG) Farmers are unable to cross, because they would not be able to set aside the seven fields necessary to cross parents and offspring (Plant breeder, MINAG) Farmers are not capable of maintaining good seed quality. They're better off going to the Seed Enterprise, and the same goes for seed saving (Seed Enterprise staff)
Potential for intercropping	Beneficial in terms of land use equivalence and positive impact on pest control. Makes better use of land space, reduces pest and disease, protects soil, higher yields (Research: ACAO, 1999) Can obtain more profit and a more regular income (Farmer) Takes best advantage of the land, saves on irrigation, and beneficial for plant interactions (Farmer)	Induces crop competition, difficult for mechanization, less ventilation, lower yields. More labour intensive so only appropriate on smaller farms (Plant breeder) Intercropping plants has negative effects and harvest times are difficult (Farmer)

Production issue	Opinions favourable to organic production	Opinions challenging organic production
Yield potential	Organic yields higher than industrialized by 7–14t/ha (Research: ACAO, 1999) Average yields have increased from 1994 to 1996 using organic methods (Farmer)	Production has decreased under low-input methods. (Farmer) Tomato yields dropped from 27t/ha to 17t/ha (Farmer)
Economics and production costs	Income remains steady due to higher value crops chosen during conversion period (Farmer) Improved – because of higher market prices, lower input costs and higher yields (Farmer) Production costs decreased on organic *fincas* while rising on industrialized *fincas* of the same cooperatives (Research: ACAO, 1999)	It takes many years to see good results (Farmer) Worse – because of lower yields, lower prices of some crops, deteriorating machinery, higher input and labour costs (Farmer) Production is lower and therefore there is a bad economic impact (Farmer)
Effectiveness of organic soil fertility strategies	Maize does not require any chemical fertilizer, except if very high yields are required (Farmer) Good maize yields can be achieved with good organic management, including addition of 12–30t/ha manure (Researcher) Soil fertility increased, due to better treatment. (Farmer)	Maize cannot benefit from organic fertilizer as it is a short cycle crop (Plant breeder) Soil fertility decreased, due to previous poor soil management and lack of fertilizers (Farmer)
Effectiveness of biological pest controls	Applying Bt at one month and frequently thereafter is sufficient control for maize if combined with heavy rains (Farmer) *Thrips* can be controlled by breaking the early life cycle with biological inputs (Plant breeder) Biological control is cheaper, better for human and soil health, easy to apply, effective (Farmer)	It is not possible to produce maize without the use of chemical biocides (Farmer) It is not possible to control *Thrips* only with biological controls (Staff, Sanidad Vegetal) We are satisfied with pesticides. In some cases there is no other way to control a pest. Pesticides are more effective and easier to apply (Farmer)

Table 9.1 – *continued*

Production issue	Opinions favourable to organic production	Opinions challenging organic production
Labour requirements	Labour requirements decrease once the ecological basis is established. (Farmer)	Farmers perceived a higher labour input. (Research: ACAO, 1999)
Product quality	Increased because of fewer chemicals. (Farmer)	Decreased, because product size has decreased. (Farmer)
Potential for scaling up organic agriculture	Good potential based on positive experiences with increased production during the 1990s, given the right conditions. (Farmer) Possible, given the right conditions and the political will. Needs care and attention. (Farmer)	Agrochemicals will continue to be used because organic agriculture is confronted with difficulties. (Plant breeder) Large-scale production and high technology are required to meet national demand. (Seed Enterprise) Yields would drop. Too much work. A risk. Soil fertility too poor. Insufficient organic inputs. (Farmer)

Source: Wright, 2005

A number of factors could account for these sometimes contradictory opinions. These factors included the differing economic situations and needs of farmers; differing value interpretations (for example, of what was meant by 'good yields' or 'quality'); differing criteria on the means to achieve the shared goal of ensuring food security, differing levels of ecological literacy; personal experiences; worldviews; and combinations of these. Importantly, the institutional support sector underestimated the real capacity and ability of farmers to successfully manage their own resources.

The degree of uptake of organic concepts related to levels and type of education. Those farmers and support professionals who had been exposed to training on organic agriculture held a higher awareness of alternative options and strategies, increased ability to use these to innovate, and were less likely to envisage dependency on chemical inputs. The converse was also seen. Based on these factors, certain actions might usefully be taken to increase levels of ecological literacy, as follows:

• Replacement of the terms 'low input' or 'resource poor' with 'appropriate' agriculture.

- Broad dissemination of existing research results that verify the benefits of organic agriculture, and replication and scaling-up of other research for further verification.
- Recognition of the potential capacities and capabilities of farmers and/or ecological systems to deliver.
- Increase in training and knowledge on the scientific basis of organic agriculture, and on the epistemological differences to industrialized approaches.
- Stimulation of debate on the meaning and interpretation of concepts such as product quality, yield optimization, soil fertility and human health, in relation to their utility for ensuring food security of the population.
- Education on the appropriate time and place for the application of industrialized techniques.

Raising ecological awareness was required not only for those working in agriculture. A university lecturer explained:

> It is impossible to attain sustainable development of society without a sustainable agricultural sector and the safe food production it produces, and vice versa. In this light it is evident that training of traditional target groups is not enough. A comprehensive quantitative determination of training needs would include almost all of the 11 million Cubans.

The need to support innovation and experimentation

As well as generating knowledge to fill some of the identified gaps, more localized farming systems required an ongoing process of locally appropriate knowledge generation. Pre-identified gaps in knowledge identified by institutional professionals included non-chemical weed management, the use of polycultures, water conservation, alternative energy on-farm, specific pests and diseases, appropriate diversification strategies, allelopathy, minimum tillage and optimum planting dates. In addition, farmers also requested new innovations such as more strategic development projects to encourage a move beyond the use of inputs of any kind, the encouragement of both diversification and regional specialization, and the development of alternative energy sources on-farm. Given the difficulties in accessing knowledge from abroad, generating knowledge in-country was particularly important, and recovering and incorporating traditional knowledge into the research process was seen as being critical. Awareness was increasing of the need for locally appropriate techniques, compared to the 'one size fits all' approach of industrialized technology transfer. Individual farmers would have to test for themselves whether, for example, it was appropriate to use oxen on soils which might be compacted or dry, whether sufficient care and attention could be provided for these animals if the farm lacked basic resources, or whether certain biological control products could be used in unfavourable weather conditions.

For on-farm learning and innovation to occur, two factors were acknowledged as being crucial: the presence of traditional knowledge, and – for non-CCS cooperatives – the practice of 'linking man to the land'. Enabling a more direct relationship, and incentive scheme, between an individual and a land area (as opposed to team work) increased the opportunity for innovation. Providing a more supportive environment for innovators would also necessitate a greater degree of flexibility within the planned production agreements.

Additionally, amongst the successful examples of innovative projects and activities, one common feature stood out. The majority had been initiated and driven by individual pioneers who held a clear vision of what was required to address a specific situation and who had initially worked against the tide of public consensus and frequently in isolation. Such 'ecological pioneers' were found on farms, in research institutes, or as founders of whole organizations. Box 9.1 has noted the work of the Cuban organic movement as an institutional example of an ecological pioneer. ACAO had made an important impact behind the scenes throughout the 1990s, but in taking a facilitatory role its impact was not so obvious to people working outside organic circles; its partners often received the credit for projects that it instigated. At the same time, ACAO had received little financial backing and was therefore limited in its scale and mode of operations. State backing came in 1999, when international demand to collaborate with Cuba on organic projects spurred MINAG to transfer ACAO's more independent role to a government-appointed entity, the Cuban Association of Agricultural and Forestry Technicians (ACTAF). However, none of ACTAF's objectives mentioned organic or ecological agriculture, rather its statutes included that 'ACTAF may work through a wide field of actions, mainly directed at balancing rational, sufficient and efficient use of chemical methods with organic and biological alternatives' (ACTAF, 1999, pp12–13). Although this institutionalization of the grassroots organic movement represented a recognition by MINAG of the potential of organic approaches at least for strengthening international relations, it ran the risk of weakening the pioneering spirit of the organic movement.

The promotion of participatory approaches also rested on the efforts of pioneering individuals, such as those who introduced the concept of participatory plant breeding into Cuba to alleviate some of the challenges of seed supply, as described in the previous chapter. These pioneers had to take personal initiative and risk to get such projects off the ground in the face of mainstream mandates and with few financial resources. Additionally, specific farmers, working either alone or on cooperatives throughout the country, were the grassroots driving force behind the spread and success of organic agriculture both on-farm and within regions. Some of these farmers were acknowledged and used as demonstration farmers by ANAP and other groups, but many worked unrecognized. One such farmer pioneer in Havana Province, for example, had pioneered the conversion of part of the CPA cooperative on which he worked to organic, and was instrumental in training farmers on agro-ecology. Box 9.2 describes the perceptions and practices of this farmer.

Earlier recognition and support for these individuals and groups would further encourage the spread of innovation and experimentation, whilst being mindful of not taking control and weakening their progressive spirit.

The need to develop and encourage appropriate methodologies

Organic techniques were frequently disseminated through the same, pre-existing extension pathways and using the same methodologies as for the predominant industrialized system. These mass transfer-of-technology methodologies were effective to a certain degree; the development and dissemination of biological inputs provided an example of this. Participatory methods, more appropriate for localized production, were also developing, but teething troubles had arisen. Based on the experiences of research and extension professionals from the Science and Technology Department of MINAG, significant insights and challenges were identified that needed to be addressed in order to improve the more widespread use of these appropriate methodologies. These insights included the retraining of professionals not only in participatory approaches but also in social sciences, the need to develop trust with the farmer, the higher success rates of working on CCS cooperatives and outside the structure of planned production, the challenge of introducing new methods into the establishment, the need for recognition of the new role of facilitator, and the need for recognition that yield indicators were insufficient to judge project success.

Increasing the availability of, and access to, appropriate resources and technology

The second major factor to enable the scaling-up of organic agriculture was the need for improved access to organic inputs such as biological pest controls and manure. This was considered crucial by both farmers and institutional professionals and especially in order to turn around the process of soil degradation. By the end of the decade, agrochemicals were becoming increasingly available and were considered by many to be easier to apply and faster acting, albeit more expensive. It was not only input access but also availability, price and delivery that were of concern to farmers. Specific inputs, resources and technologies

Box 9.2 *Perceptions and practices of organic agriculture by a pioneer farmer veteran in Havana Province*

CPA farmer Ricardo Manuel had first encountered organic agriculture back in 1992 at a seminar heralding the inception of the Cuban organic movement. This approach was not new to Ricardo; it was similar to traditional agriculture that farmers had practised for centuries, whereas he had been encouraged to farm industrially for only four and a half decades. For Ricardo, organic agriculture was about timely planting, field rotations, soil improvement, minimal labour, crop associations, polycultures and enhancement of biodiversity. For him also, industrialized agriculture did not make good economic sense. He explained why he had had enough of this form of agriculture in the context of the Special Period: 'When the camel is in the desert with a long walk and a heavy load, he asks the flea to get off his back.' Why did he have this belief in organic agriculture when other farmers did not? Ricardo responded that 'Those who can believe are those with the most education and knowledge,' and added that 'People change when things start to work against the reality of what they see as being true.'

Ricardo had converted one of the *fincas* of the CPA to organic, and found that although bio-fertilizers were useful for the transition process they were unnecessary as part of a good ecological system. They were costly, and he found a better alternative to cover the shortfall in the low-yield conversion period was to grow higher-value crops. Owing to his use of green manures and minimum tillage, he had raised soil organic matter content from 0.5 in 1993–94 to pre-industrial levels of 5.0–6.0 by 1999. For pest and disease control, Ricardo did not feel that agrochemicals were completely redundant, but that they tended to focus on the wrong target, and although he had used biological controls during the conversion period, by 1999 these were no longer needed. Yields were now starting to increase as were biodiversity levels, and with rising bird numbers so there were fewer pest outbreaks. Although total yields were increasing, Ricardo pointed out the four-year, cyclical fluctuation in yields that related to weather patterns. To mitigate this cycle, in the difficult years he planted more resistant crops such as sweetpotato and cassava.

More labour had been necessary at the early stage of the conversion, but once the ecological basis was established, after 5–6 years, then labour requirements decreased. Ricardo stressed the importance of the learning process and the time required for this to occur. His farm colleagues were still noticing the differences in quality as regards food flavour and colour, as well as seeing the cost savings of the organic *finca*. Although Ricardo had never been in formal education, he had learned by 'reading, listening and looking', and now ran training courses on agro-ecology: 'Agronomists especially have to start learning everything through observation: mixing plants of different colours and smells to regulate insects, noticing air currents at different times of the year and wind direction, and the influence of the sun and the moon.' For him, there was a continual learning process between the farmer and nature, and between farmers.

Ricardo felt that integration was the key to sustainable agriculture in Cuba, and health the most important consideration with a balanced production–consumption chain and a balanced diet. His country was forging its own path, based on experience and knowledge, a path that could not be planned as with the industrialized model: 'This requires a slow, step-by-step approach, and the country would be wrong to hope otherwise.'

Source: Wright, 2005

were identified as necessary in order to increase the use of organic inputs, as follows:

- The timely availability of biological pest and disease control inputs, and improved quality of these inputs including improvements in storage times.
- Availability and quality of green manure seeds.
- Appropriate crop varieties and quality seed.
- Simple receptacles for collecting biological inputs from the CREEs.
- Transport/fuel for travelling to collect inputs, from the CREEs, research centres, livestock farms and so on, and to facilitate farmer seed exchange or crop–livestock integration activities between farms, or to enable on-farm delivery.
- Finances to purchase those inputs that had risen in price, particularly manure, or regulation to maintain low price levels of these inputs.
- Fodder for oxen.
- Traditional artisans such as blacksmiths, yoke producers and ox-drawn implement makers.
- Refrigerated storage for conserving soil activators such as *Azobacter.*
- A formalized institutional network for the diffusion of soil fertility inputs, based on the model of the CREEs.[3]

Ensuring conducive socio-political factors

Ensuring policy support

In many respects the political responses to the crisis of the early 1990s favoured an organic approach, particularly the emphasis on self-reliance. Yet other elements of the political imperative served to constrain the widespread uptake of organic practices, especially towards the end of the decade. These five elements relate to legislation, yields, energy sources, planning approaches and quality considerations.

Environmental legislation > agricultural legislation
Agricultural policy was vague; it referred to concepts such as integration and sustainability but without further definition. Specific legislation on agricultural contamination did exist but was contained in environmental laws emerging from CITMA rather than from MINAG.[4] Professionals in the agricultural sector indicated the need for tighter enforcement of existing environmental legislation on issues such as soil protection, and for increases in penalties payable for the misuse of agrochemicals. In fact, CITMA and MES were supporting more organically oriented projects than MINAG. Given Cuba's international

reputation for leading on organic agricultural policy, the development of more progressive agricultural legislation would lead the way for other nations as well as increase the efficacy of its own agriculture.

Maximizing yields > optimizing yields

The nationwide objective for a rapid increase in yields directly conflicted with sustainability objectives. It could be inferred as against government policy to promote strategies that might result in lower yields, such as based on the organic perspective of optimizing rather than maximizing yields. By the end of the decade, research results showed the frequently superior performance of organic production, yet these results were not being taken into account when formulating agricultural policy.

Lack of petrol > renewable energies

The pervading lack of petrol held back the development of organic agriculture in its input-substitution state. Cuban farmers saw petrol as being pivotal to the success of their production systems, to fuel both transport and irrigation. Drought-resistant crop varieties, for example, were measured not so much by their yield performance but by the fuel savings that would be made through the decreased need to operate irrigation pumps. Unless organic alternatives could address farmers' concerns of irrigation and traction limitations, they were less likely to be taken up. Efforts to pilot renewable energies were driven largely by one small NGO; feasibility studies indicated that Cuba could meet all its electricity needs from sugarcane bagasse. At the end of the decade, however, progress on alternative energy was negligible and continued to rely on international donor support (Montanaro, 2000), with little state funding prioritized for investing in renewables on a large scale.

Industrialized planning > ecological planning

Several aspects of agricultural and food planning and organization worked against organic agriculture. The state production plan was based on tonnage and advocated crops that were not always appropriate to the locality. These plans could include the provision of a technology package of agrochemical inputs. The national farm intensification programme worked against the use of green legumes in rotations, intercropping and fallows. In the seed sector, the centralized seed distribution system worked against farmers developing their own skills and expertise for locally appropriate varieties. To reverse these impacts would require a planning system grounded in an ecological perspective.

Quantity incentive > quality incentive

Food quantities were still unstable, and price and quantity were the most important factors in the market place. The majority of crops that fed into the production plan were not differentiated by quality, and nor was there any financial stimulus for this (except for the tourist market). Any quality differentiation was again based on product size and weight, and secondarily on pest residues. The ration system also limited the opportunity for the development of farmer–consumer relations over quality produce. Far from detracting from good and consistent yield performance, a focus on quality would also bring benefits in terms of residue-free and nutritionally high value produce. Such a refocusing could also be used as a vehicle around which to raise awareness on human health and nutrition issues, as well as on farming techniques that improved the quality of the natural resource base.

Supporting social sustainability and change

Common social challenges

Certain social factors were also crucial for the scaling-up of organic agriculture, and two social disincentives in particular required addressing. Some farmers were unwilling to adopt technologies and practices that they felt were unproven or that were not being diffused by the state. For example, at the end of the decade, some farmers indicated that they would wait for the state to re-introduce biological pest controls rather than attempt to obtain them directly from the CREEs. This attitude was a remnant of the previous top-down extension system that had, to some degree, inculcated dependency and even mistrust. The other social disincentive was the incidence of theft from fields and stables, which limited farmers' crop choices, their seed drying and saving activities, and keeping oxen on-farm if a guard could not be afforded. These social issues of theft, mistrust and dependency would require some time to turn around.

Social recognition of the need for change

A recurrent theme was that practical change required a corresponding shift in mentality, and that 'attitudes take time to change'. This viewpoint, shared by many, tended to inhibit any attempt to encourage change, because of the anticipated negative response. At the same time, professionals in the agricultural sector had no difficulty in identifying the need for a mental shift by some other group or individual, but never for the professional him- or herself. Perhaps because of this, the provision of training in organic agriculture was targeted largely at farmers rather than support staff, yet in reality many farmers were further down the transition pathway. Given the directive approach of the institutional sector in Cuba, re-training and attitude change amongst agricultural professionals were critical for a more widespread change. Following this, supporters of organic agriculture in Cuba were at the end of the decade emphasizing the need for a shift in consciousness, in order to move from the stage of input substitution to that of ecological management. Treto et al (2002, p184) suggested that 'This will require a deep-rooted paradigm shift, already under- way, allowing agronomists and farmers to view the soil as a living system of an agricultural ecosystem that operates according to the laws of nature.' Similarly, Ríos Labrada et al (2000a) recommended that whilst significant technological change towards organic agriculture has already occurred over a relatively short timescale, this needed to be accompanied by a campaign of environmental and health education. The authors concluded (p16): 'It is more difficult to raise con- science than technology, and this is the challenge for the future.'

Potential driving forces for the scaling-up of organic agriculture

Cuba's successes in recovering its food security status and agricultural produc- tivity suggested that, if the political will was there, the country also had the capacity to develop joined-up policy measures and an enabling environment to support a more sustainable agriculture. Such a move would involve not only the strengthening of existing organic strategies, but also modifying those others that conflicted with this goal.

The motivation to scale up organic agriculture could come about for sev- eral reasons. In a period of just ten years, Cuba had made a transition from facing serious food deficits and shortages in calorific intake, to one in which more than one-third of the population of Havana was considered to be over- weight and Western diseases prevalent. For although the state had solved some of the problems of inadequacy of food supply, its focus on quantity over quality issues appeared to be somewhat counter-productive. Cuba was showing that it was possible to ensure food supplies no matter whether its domestic production system was industrialized or more organic, but the predominant agricultural strategy had broader and more serious consequences for the health of the nation, and of the environment. Soil degradation remained a huge and restric- tive problem for the agricultural sector, as did the recurrent droughts, which required more adaptive, dynamic and resistant crops and cropping patterns, and water management systems. As was pointed out by one agricultural spe- cialist, 'The tropical climate of an island like Cuba is inherently unstable, and on a large-scale model, the farmer cannot react in time to the change in climate' (quoted by Sinclair and Thompson, 2001, p27). At the same time, the visible benefits of organic production included the positive human health impacts of reducing agrochemical use, and the food system was benefiting from a more diverse range of fresh foods. Other potential benefits of organic production, such as supplying an organic export sector or growing internal tourist market, might also play a motivational role.

There were other more fundamental reasons why Cuba in particular might wish to consider the organic option. Industrialized agriculture had been developed by and for temperate agroclimatic regions, where relatively large, flat land areas, stable soils, even rainfall, lower levels of biodiversity and sunlight were able to support it. Industrialized agriculture was far less compatible with tropical conditions (Weischet and Caviedes, 1993). As one Cuban NGO staff explained, 'The country is producing more now than it was ten years ago – industrialized agriculture never worked here properly – even when supported by the USSR – it only produced less than 50 per cent of food needs.'

In terms of strategy, a short-term focus on yield maximization was appropriate in emergency situations when a rapid response was required. By 2000, Cuba was moving out of the crisis and able to cast a glance towards more appropriate longer-term approaches. Industrialized systems were also possibly incompatible with a country that prioritized societal interests over fiscal accumulation. In a speech entitled 'Cuba's Green Path', the then Minister of CITMA explained: 'The profit motive plays only a limited role in Cuban agriculture, healthcare and energy programmes. As these programs are largely state-supported, society-wide needs of sustainability can take precedence over profit margins' (Montanaro, 2000, p1). One agricultural researcher put it another way: 'Lenin went wrong on copying the industrialized agricultural model of capitalist countries – it does not work, especially centralization.' Castro (1993, p24) recognized this paradox and interpreted it from the environmental perspective, pointing out: 'If the deterioration of the environment is analysed from a historical perspective, it can be appreciated that the greatest harm to the global ecosystem has been done by the development patterns followed by the most industrialized countries... The North's ecological deterioration has been exported – in large measure to the South, as part of a long process of capitalist development.'

Notes

1   The Latin American school is described by Altieri (1995): 'Agro-ecology provides the basic principles to study, design and manage alternative agricultural ecosystems. It takes into account both the ecological and environmental dimension, and the economic, social and cultural aspects of the crisis of modern agriculture' (cited in Monzote et al, 2002, p208).

2   Pérez and Vázquez (2002, p136) explain in more detail: 'Everyone asks what will become of ecological pest management in Cuba, as we emerge from the economic crisis of the early 1990s. As more foreign exchange becomes available for the purchase of pesticides on the international market, it seems logical to some that Cuba will return to an intensive dependence on chemical inputs. Moreover, some think the current programme of accelerated reduction of pesticide use is simply a short-term,

stop-gap answer to maintain production until pesticide imports are affordable once again. But others – and they are more than a few – have a very different analysis, looking seriously at economic, social, health, and environmental factors, and conclude that the agro-ecological IPM model developed to date is simply a better model.'

3　As Treto et al describe (2002, p179), 'In spite of all the advances in the production and application of worm humus, bio-earth, compost, *cachaza*, and bio-fertilizers, we continue to have difficulties in substituting for all the chemical fertilizers. Often not enough raw materials are available when and where they are most needed. Moreover, taking a macro view, there simply are not enough of these materials to provide soil amendments for all agricultural production at a national level. Even if there were, transportation of such large quantities of material is costly and logistically complicated. On-farm solutions will have to play an increasing role.'

4　As in many other countries, concerns about the environmental and health aspects of agriculture have been seen as separate from those related to production, and are dealt with by separate ministries.

References

ACAO (1999) *Informe Final de Proyecto SANE*, Habana (unpublished)

ACTAF (1999) *Convocatoria, Tesis, Estatutos y Reglamentos*, La Habana, Primero Congreso, ACTAF

Altieri, M. (1995) *Agroecology*, second edition, London, IT Publications

Casanova, A., Hernández, A. and Quintero, P. (2002) 'Intercropping in Cuba', in Funes, F., García, L., Bourque, M., Pérez, N. and Rosset, P. (eds) *Sustainable Agriculture and Resistance: Transforming Food Production in Cuba*, Oakland CA, Food First Books, pp144–154

Castro, F. (1993) *Tomorrow is Too Late. Development and the Environmental Crisis in the Third World*, Melbourne, Ocean Press

Castro, F. (1996) Speech at UN World Food Summit, 18 November

Funes, F. (2002) 'The organic farming movement in Cuba', in Funes, F., García, L., Bourque, M., Pérez, N. and Rosset, P. (eds) *Sustainable Agriculture and Resistance: Transforming Food Production in Cuba*, Oakland CA, Food First Books, pp1–26

Garcia, M. (1998) *Contribución al studio y utilización de los abonos verdes en cultivos económicos desarrollados sobre un suelo ferralítico rojo de La Habana.* Unpublished PhD dissertation, UNAH, Havana

Garcia Ramirez, M. J. (1999) *Valoración Biológica del Uso de las Micorrizas Vesiculo-Arbusculares en dos Patrones de Citricos.* Tesis de MSc, Instituto de Ecologia y Sistematica, Ministerio de Ciencia, Tecnología y Medioambiente, Havana

Instituto de Suelos (1998) *Bordos de Desague, Una Tecnología para Reducir las Pérdidas del Suelo. Boletín No. 1*, Havana, MINAG

Martínez Viera, R. and Hernández, G. (1995) 'Los biofertilizantes en la agricultura cubana', *Resúmenes del III Encuentro Nacional del Agricultura Orgánica, Conferencias*, Havana, p43

Maura, J. A. (1994) 'Producción de biopesticidas. El caso de Cuba', *Informe del Taller Regional sobre Tecnologías integradas de producción y protección de hortalizas*, Cuernavaca, Mexico, FAO, pp69–74

Montanaro, P. (2000) *Cuba's Green Path: An Overview of Cuba's Environmental Policy and Programs and the Potential for Involvement of US NGOs*, California, Cuba Program, Global Exchange. www.globalexchange.org

Monzote, M., Muñoz, E. and Funes Monzote, F. (2002) 'The integration of crops and livestock', in Funes, F., García, L., Bourque, M., Pérez, N. and Rosset, P. (eds) *Sustainable Agriculture and Resistance: Transforming Food Production in Cuba*, Oakland CA, Food First Books, pp190–211

Orellana, R., Dibut, B., Fundora, Z. and García, D. (1999) *Agricutura Organica: una definition absoluta?* Havana, IIFAT, ACTAF

Paredes, E. (1999) *Manejo agroecológico de malezas y otras plagas de importancia económica en la agricultura tropical*. Curso sobre bases agroecológicas para el MIP, Matanzas, Cuba

Pérez, N. and Vázquez, L. L. (2002) 'Ecological pest management', in Funes, F., García, L., Bourque, M., Pérez, N. and Rosset, P. (eds) *Sustainable Agriculture and Resistance: Transforming Food Production in Cuba*, Oakland CA, Food First Books, pp109–143

Pesticides Trust (1998) 'Cuba's organic revolution', *Eco-Notes*. www.ru.org/81econot.html

Pozo, J. L., Arozarena, N., Carrión, M., Gonzalez, R., Companioni, N. and Rodríguez, J. (2000) *Propuesta de manejo agrosostenible para el cultivo de maíz (Zea mays L.)*, INIFAT, Cuba. IV Organic Conference on Organic Agriculture, Havana, May

Quintero, P. L. (1999) *Evaluación de algunas asociaciones de cultivos en la Cooperativa Gilberto León de la provincia La Habana,* Havana, Universidad Agraria de La Habana, CEAS

Riera, M., Mendez, M. and Medina, N. (1998) 'Uso de biofertilizantes en secuencias de cultivos y sus influencia en el sistema del suelo', *Proceedings, VI National Conference on Sustainable Agriculture*, INCA, Havana, 22–25 November

Ríos Labrada, H. and Wright, J. (1999) 'Early attempts at stimulating seed flows in Cuba', *Magazine of the Information Centre for Low External Input and Sustainable Agriculture (ILEIA)*, vol 15, no 3/4 (December), pp38–39

Ríos Labrada, H., Funes, F. and Funes Monzote, F. (2000a) *Alternative Food Production in Cuba: Strategies, Results and Challenges*. Unpublished paper, Havana

Ríos Labrada, H., Almekinders, C., Verde, G., Ortiz, R. and Lafont, P. R. (2000b) *Informal Sector saves Variability and Yields in Maize. The Experience of Cuba* (unpublished manuscript)

Ríos Labrada, H., Soleri, D. and Cleveland, D. (2001) 'Conceptual changes in Cuban plant breeding in response to a national socioeconomic crisis: the example of pumpkins', in *Farmers, Scientists and Plant Breeding. Integrating Knowledge and Practice*, Wallingford, CAB International

Scialabba, N. (2000) 'Factors influencing organic agricultural policies with a focus on developing countries', *Proceedings, IFOAM Scientific Conference*, Basel, Switzerland, 28–31 August

Sinclair, M. and Thompson, M. (2001) *Cuba Going Against the Grain: Agricultural Crisis and Transformation*, Oxfam America. http://www. oxfamamerica.org/newsandpublications/publications/research_reports/art1164.html accessed May 2002

Treto, E., Garcia, M., Martínez Viera, R. and Manuel Febles, J. (2002) 'Advances in organic soil management', in Funes, F., García, L., Bourque, M., Pérez, N. and Rosset, P. (eds) *Sustainable Agriculture and Resistance: Transforming Food Production in Cuba*, Oakland CA, Food First Books, pp164–189

Weischet, W. and Caviedes, C. N. (1993) *The Persisting Ecological Constraints of Tropical Agriculture*, UK, Longman Group

Wright, J. (2005) *Falta Petroleo. Perspectives on the Emergence of a More Ecological Farming and Food System in Post-Crisis Cuba*. Doctoral thesis, Wageningen University

10

Lessons for the Future:
Cuba Ten Years On

Cuba's successes and challenges in meeting its food security goals

Cuba's successes: a united social front

The strategies put into place by the Cuban government to cope with the crisis are taken for granted from within the country, but from an outsider's perspective they provide a unique global example of a different socio-political approach to improving food security. Through an innovative and pragmatic mix of measures and mechanisms in its agricultural and food sectors, Cuba managed to turn around a serious and nation-threatening crisis. It reduced its dependence on imported food and inputs and, although food aid and subsidized imports were still necessary, the country moved closer to meeting national food needs domestically over the longer term, all the while maintaining an equitable social food programme under extreme resource-limiting conditions. Major changes put in place by the state, or developed by lack of choice, included:

- development and maintenance of human and social capital
- joined-up political will across sectors
- development of subsidiarized food action plans
- partially controlled food distribution and social safety nets
- mandate for local, regional and national self-sufficiency in production
- substitution with organic production techniques: biological pest and disease control and fertilizers, farm diversification, rotations and intercropping, increased use of draught power, and training in agro-ecology
- enabling of land access from household to farm level
- development of a strong urban agriculture movement
- encouragement of urban to rural migration
- diversification of accessible markets and food sources
- raising of farm gate prices
- shortening of market chains
- downsizing of large farms
- reconnection of farmers and workers to land

- government investment in agricultural research, extension and training
- increased postharvest efficiency
- national autonomy over food imports, exports and international relations.

The cohesive political will cannot be underestimated. Cuba had a strong government committed to overcoming the crisis and maintaining its socialist principles. This was exemplified by the national mitigation strategy, 'The Special Period in Peace Time'. The government was able to mobilize broad support and cooperation between different ministries. Those ministerial institutions involved in agriculture, food distribution, economics and health were obvious examples, but even the military was called upon to help out. This political will extended not only horizontally but also vertically from national, through provincial to municipal levels, and the communication and hierarchical staffing structures were such that policy could be implemented nationwide, in the fastest possible manner. To the extent possible, everyone had the same mandate and acted in unison. There was no unregulated private sector or social promotion of self-interest that might contradict and dissolute efforts. In fact, for those visiting Cuba from neo-liberal countries, the relatively high level of community concern amongst the population was striking.

On the one hand, centralized planning was crucial in order to translate increases in production into adequate food access, which was considered a public right or good. As in other food-insecure countries, food supplies were unstable for a number of reasons, yet the planned food collection and distribution system took the responsibility to collect rural surpluses and transport these to urban areas, and to ensure the transportation of imported staple foods to remote rural areas. On the other hand, the state gave up some degree of control over management of food production and distribution, and encouraged initiative and activity at the grassroots level. This was achieved, for example, through the transformation of state farms into UBPCs, and also by introducing market mechanisms while keeping out a watchful, regulatory eye to ensure that these markets did not lead to excessive inequality. The combination of a guaranteed basic income, buoyant farm gate prices, shortfalls in supply and regulated markets protected farmers from getting trapped on the productivist treadmill (which in other countries sees farms going out of business on a daily basis). Further, and contrary to neo-liberal economic theory, when farmers had the choice they did not always opt for those markets with the highest financial returns. Other factors influencing their market decisions included the ease of transport to market, labour availability and guaranteed prices offered by the state.

In this sense, the state shifted its role, at least partially, from provider to facilitator, and food production and distribution were moving towards a relatively localized pattern with the emerging realization that the 'one size fits all' approach of industrialization had not delivered. Localized production reduced transport costs, eliminated external food dependency, and reduced postharvest physical and nutritional losses. On the other side of the farm gate, localized production – the human-scale reconnecting of farm workers with the land and corresponding micro-management – was a more critical factor in raising yields than, for example, the use of organic inputs (Rosset, 1996). The superior resilience and productivity displayed by *campesino* farmers in the early 1990s was recognized to the extent that elements of their production systems were taken and applied to other forms of farming cooperative. Organic approaches played a role and were demonstrated as technically feasible components of a national food security system. By the end of the decade, Cuba held more relative sovereignty over its food system than at any time in its recent history. The long-term investments in social infrastructure and human resource base formed the bedrock for the resilience displayed throughout the 1990s, and this was coupled with the state's capacity for rapid learning and change, which included, for example, the move towards a more market-led approach, and measures to wean the population off the ration system while ensuring that this did not impact on the poorest.

Although Cuba is fairly unique in its mode of governance as compared to other countries experiencing food insecurity, its success may be measured against the reforms made by other former Socialist Bloc countries after 1989. Cuba's cautious experiments with market mechanisms compare favourably with the more rapid and widespread adoption of liberalized market approaches of post-Soviet Eastern European countries and their impacts on agriculture and rural livelihoods. Box 10.1 summarizes the somewhat scarce literature on this issue.

Cuba's agriculture has also been compared to that of a Caribbean neighbour, the Dominican Republic (Sinclair and Thompson, 2001). In the Dominican Republic, agriculture is characterized by deregulation, large-scale farms and foreign ownership. At 2000, and not surprisingly, Cuba's agriculture showed higher social indicator levels (such as equity and employment) and lower productivity levels. With regard to the trade embargo imposed by the United States, the measures taken by Cuba were not the only options available but were comparatively successful. Other countries responded differently to restrictive embargoes. For example, during the embargo faced by Iraq in the 1990s, the Iraqi government survived through corruption, and prioritized the well-being of the military over Iraqi civilians who suffered from disease and malnutrition. In the late 1980s, South Africa coped with its oil embargo by turning to coal to power the economy, which led to short-term growth and continued a dependency on non-renewable energy (Wiskind, 2007).

> **Box 10.1** *The impacts of the rapid introduction of liberalized markets on agriculture and rural livelihoods in post-Soviet Eastern Europe*
>
> Prior to 1989, socialist agriculture had failed to meet planned targets, thereby giving the impression of a sector in crisis. However, underlying this had been substantial investments in buildings, machinery, fertilizers and irrigation systems, which provided food for the population at affordable prices (Turnock, 1996). After the collapse of the Soviet Union and the end of price controls and trade contacts, the transition to a market economy cast agriculture into a state of turmoil. As wages fell, so domestic demand for food products was reduced. Overtures made to the European Union to become a major trading partner revealed that this market already had substantial food surpluses. In contrast with the approach taken by Cuba, the economic systems of other former Socialist Bloc countries shifted directly from state-owned, centrally planned and directed systems to market-oriented private enterprises. The agricultural sector faced major problems during the transition period as it lacked the knowledge and experience required to organize, finance and manage private enterprises (Csizinszky, 2003). Nationalized land was returned to owners who lacked the skills and financial resources for intensive crop production, and this resulted in reductions in crop yields. With a decrease in investment for agricultural research and extension, by 1998 research systems were still in the process of transition and were lagging behind reforms in the agricultural sector (Csaki, 1998). Estimates gave until 2020 for Eastern Europe to become a significant exporter of agricultural commodities (Antonelli, 1990; Molnar, 2003). More recently, efforts were put to designing new, low-input and organic cropping systems and conservation tillage.
>
> These changes had profound social impacts. A household survey of the economic and psychological well-being in rural Russian villages following the collapse of the Soviet Union, showed that, while the introduction of more competitive economic development programmes had improved the material quality of life for certain households, it had brought higher levels of stress that were exacerbated by the loss of social support previously provided by collective organization (Patsiorkovski and O'Brien, 1997). Overall, there had been an alarming decline in public health. Life expectancy decreased in many countries and the trend was forecast to continue; in Russia, male life expectancy dropped by six years between 1989 and 1994. Medical causes of mortality among adults included cardiovascular disease, cancer and injuries, but were underpinned by a sharp increase in poverty, social disintegration and crime, combined with historically high rates of smoking, alcohol use and psycho-social stress (Little, 1998).

Cuba's challenges: the health imperative

Despite the sustained increases in production and food availability, ten years was too short a period to rebalance the substantial deficit in supply or to end the dependency on imported foodstuffs. The new UBPC cooperatives confronted several teething problems in their development, and cooperatives in general lacked training in the new types of marketing channels on offer. Owing to the paternal legacy, as well as to the substitutive nature of industrial agriculture, farmers continued to look for off-farm inputs and wait for advice or instruction, rather than taking initiative themselves.[1] The general lack of fossil fuel continued to hinder the development of the whole sector, from storage facilities to transport for on-farm research. Fluctuations caused by natural hazards such as cyclones and droughts, and inadequate postharvest measures (particularly storage), remain major barriers to the food security goal. Notwithstanding the equitable access to food, some social disparities were widening, at least partly because of the ability of a portion of the population to access dollar remittances from abroad. Another factor emerging as an influence on inequality was the divergence in quality of food sold to tourists compared to that provided to Cubans. Food quality was not based purely on size and appearance; fewer chemicals were used on foods destined for the tourist market, based on the understanding that tourists enjoyed 'healthy products'. Similarly, there were plans for genetically modified crops to be grown for the Cuban population but not for tourist sales. This divergence appeared at odds with the nation's socialist concern for human health.

Although nutritionists were aware of the link between agriculture and nutrition, those in the agricultural sector were less so, and the push to increase output had overridden quality and nutritional considerations. During the crisis, stated one health official, 'It was more important to feed the people than to eliminate contaminants.' The small amount of evidence available indicated the possibility of high chemical residues in certain prioritized crops. Yet the scarcity of agrochemicals limited research into their toxicity, and the privatization of state farms made it more difficult to monitor and control agrochemical usage. Shortfalls in the nutritional adequacy of diets were dealt with through the provision of supplements rather than by addressing more fundamental issues such as soil health or crop nutrient supply. Meanwhile, the natural resource base and particularly soils were still suffering the industrialized impacts of degradation.

A common perception encountered amongst both farm workers and urban householders was of continued food insecurity, and lack of access to self-provisioning and irregular incomes appeared to be contributory factors to this. In reality, most people had access to, and consumed sufficient, if not excessive, quantities of carbohydrates, sugars and fats. This insecurity was more understandable for those people not living on farms, for whom sources and supplies of food could be precarious – the daily distribution of the ration meant that households were unable to build up food reserves. The number of undernourished people in Cuba continued to increase slightly up to the end of the decade, although the growth rate was declining and was far lower than at the start of the decade. The lingering feeling of food insecurity of the population, coupled with the continued preference for a traditional diet heavy in carbohydrates, sugars and fats, meant that Western diseases were prevalent and obesity was on the increase. This arguably undermined the advances made in making healthy, fresh produce more widely available. Fluctuations in overweight and obesity levels mirrored dietary trends, rising to 36.5 per cent in Havana Province, as food availability continued to increase at the end of the decade. Between 1999 and 2001, sugar and its products comprised 21 per cent of total dietary energy needs (FAO, 2004). Fresh vegetables and wholegrain foods, although more widely available, were not popular food items. Despite the government's belated attempts to address this, the traditional food culture prevailed. This (re)emergence in Cuba of human and environmental health-related problems (particularly of the so-called Western diseases) illustrated how stepping up a more holistic approach to the farming and food system might provide a more efficient option than a production-focused industrial one which, in overcoming one set of problems, simply created another.

Trends into the decade of the 2000s

Trends in fuel and food imports

The experiences of Cuba as it coped to resolve its food and farming challenges have been evaluated from the early 1990s up until the years 1999–2000. This period covered the most critical time, a time when extreme circumstances could cause paradigm shifts and radical change. It forms a unique snapshot, unlikely to be repeated at least in quite the same way. Even at the end of that decade, and certainly into the 2000s, the deficiencies and isolation that Cuba had found itself in were slowly reversing their trends, and with this, many of the coping strategies of the 1990s were relaxing and even reverting. Into the new millennium, the National Statistics Office in Cuba was publishing a fuller range of informative data than had been available in the 1990s, including more recent trends in the food, farming and energy sectors between 2001 and 2006 (ONE, 2006).

Between the start of the new decade and 2006, GDP rose from $1.6 million to $2.8 million. The main growth activities were construction, mining, tourism and social systems. Cuba's international trade doubled in value and it was now trading with most countries – even with the United States with whom the value of imports rose from $220,000 in 2001 to $24 million in 2006. By far the most significant trading partner was Venezuela. The main European trading partners were Spain, Russia, Italy, the Netherlands and France, and in Asia: China and Japan. Europe, and especially Russia, was the largest export destination, whilst the majority of imported goods came from South America. The value of food imports almost doubled from $38 million in 2001 to $63 million in 2006 (another source (IICS, 2008) puts the food import value at $1.25 billion by 2007).

Over the same period, fuel and lubricant imports increased in value from $49 million to $114 million, although supplies fluctuated by year and were, at their highest, not more than double the low figures of the early 1990s. In 2005, Cuba was importing 5.1Mt petroleum and derivatives for energy, and 2.9Mt petroleum products. The value of chemical fertilizer imports increased from $1.4 million to $2.6 million, comprising urea which more than doubled in quantity from 21,666t to 48,956t. Increases were also seen in superphosphate, and potassium chloride and sulphate. Imports of chemical herbicides, pesticides and plant growth regulators increased in quantity from 37,177t to 50,281t, and of insecticides and fungicides from 2622t to 3552t. Imports of agricultural machinery varied throughout the period; almost 5000 tractors were imported in 2003, though this figure fell to 182 by 2006. Ploughing equipment rose from 256 in 2003 to 5605 in 2006. Imports of public transport vehicles numbered 534 in 2003 and 1075 in 2006. In terms of food imports, quantities of meat, milk, cereals, rice and maize all increased by 50–300 per cent. The value of livestock feed imports rose by 600 per cent to $2.7 million. Of the main export products, minerals increased significantly over the period, whilst sugarcane and agricultural goods fell by approximately 50 per cent. Tobacco exports remained fairly static.

Trends in domestic environment and energy investment

State investment in environmental protection (water, soils, air, forests, solid wastes) remained fairly static in the 2000s, at around $11 million per year. Progress has been seen in the water sector, where pollution levels dropped by 27 per cent between 2001 and 2006. Yet the productive resource base was degraded to severe levels. By 2001, 65 per cent of soils were classified as of low and very low productivity. The main limiting factor was low organic matter content, which affected almost 70 per cent of soils, and also erosion, poor drainage, acidity and low water-retention capacity, which affected 40–45 per cent of soils. Salinity, compaction and desertification also played a role. In terms of air pollution, greenhouse gas emissions dropped from 37,485Gg in 1990 to 24,733Gg in 1994, but since then have risen to 27,317Gg by the early 2000s.

By the new decade, Cuba had had sufficient time to develop a national renewable energy system, yet between 2001 and 2006 renewable energy supplies fell by 80 per cent. This fall was primarily a result of the drastic decrease in availability of biomass for use as fuel – sugarcane waste, wood, sawdust and crop wastes – which fell in quantity from 12,394t to 4811t. Some types of renewable energy installation did increase during this period – such as solar panel systems, which doubled in number from 3300 to 6200. Other types of installation remained constant. Although a small number of organizations were experimenting with alternative energy in the country, there was little national investment for their scaling-up. This could be partly attributed to the poor financial conditions of the country, but was also affected by the increasing access to oil imports as relations with Venezuela further strengthened.

Cuba had invested in and increased other forms of domestic energy production, notably its own oil reserves, which increased in production from 1.5Mt in 1995 to a high of 3.6Mt in 2002. This figure was tailing off to 2.9Mt by 2006. In the 2000s, domestic supplies of natural gas also increased, while hydroelectric power and electricity production remained fairly static. A national programme was under way to revive electric power stations that run on diesel and fuel oil, which were deemed to be more fuel efficient (*Granma*, 2007). Sugarcane by-products fell to a quarter of previous figures, reflecting the decline in sugarcane production, and this may explain why Cuba was not taking advantage of ethanol as a fuel source. Oil availability in the country was fairly stable throughout the period of 2001 to 2006, at around 2.2Mt per year. Within the farming sector, oil consumption dropped slightly over the period, from 0.2Mt in 2001 to 0.17Mt in 2006. (A similar drop was seen in most sectors apart from the social system, where consumption rose.) Domestic production of nitrogen fertilizer fell by approximately 28 per cent, and production of insecticides and herbicides fell 83 per cent.

Trends in domestic agriculture and food supplies

State investment in agriculture rose by 44 per cent from $5.9 million in 2001 to $13 million in 2006. Within this, sugarcane production fell to one-third of previous levels, and shifted from state to private production. The land given over to sugarcane, which had amounted to 1.3 million ha in the early 1980s, had remained fairly constant during the 1990s, but at around 1998 started to fall dramatically to only 397,000ha by 2006. Yields fell by more than half.

For other crops, productive areas reduced slightly, apart from fruit and horticultural production. Yields and production quantities increased over the period for most crops, particularly sweetpotato, taro, banana, some fruits and horticultural crops, although 2006 appeared to be a poor year all round, and this could be attributed to the prevailing drought and fluctuations in oil availability. Comparing state to non-state farms, yields were fairly similar with a few exceptions: state farms were achieving higher recorded yields in banana, roots and tubers, horticultural crops (*organoponicos* fell into the state bracket) and citrus, and non-state in coffee and other fruits. In the livestock sector, milk production fell by one-third over the period, largely because of the fall in livestock numbers rather than yields. Pork production increased slightly, as did the production of sheep and goats and their milk. Poultry meat fell by two-thirds, but egg production increased by approximately 25 per cent. In the fishing industry, total catch fell from 80,300t in 2001 to 54,800t in 2006. Within this, aquaculture production and shellfish catches maintained their levels and even rose for species such as farmed catfish and prawn culture.

In the food processing industry, the quantity of many products rose significantly, by 30–100 per cent, such as preserved pork, pork fat, milk powder, cheese, yogurt, vegetable oil, some types of fish product, bread and biscuits. Production quantities of other products fell, such as of beef, poultry, milk, pasta, and conserved fruits and vegetables. Sales of goods in food service establishments increased by 17 per cent, and sales of food products elsewhere increased by 25 per cent. In terms of human health, reported cases of food poisoning tripled, from 6600 in 2001 to 19,500 in 2006. The incidence of diabetes increased by 25 per cent.

By the end of 2006, state farms numbered 6000 and covered 35 per cent of the cultivable land area, a figure similar to the previous decade. Non-state farms numbered almost 5000, and of these, 57 per cent were UBPC farms, which covered 37 per cent of the land area. CPA farms numbered 690, covering 9 per cent of land, and CCS and individual farmers totalled 1400 and covered the remaining 18 per cent of land area. These figures indicate that, since the late 1990s, UBPC farms had increased in numbers and land coverage whilst CPA and CCS cooperatives had decreased. That UBPCs were emerging as the dominant form of production gives credit to the ex-state farm workers who had not only survived the transition to privatization but were apparently consolidating their position, albeit with strong state support. State farms contained a high proportion of pastures, degraded land and forest, while CCS farms had more land given over to mixed crops, coffee and tobacco.

During the period, the number of people employed in agriculture remained fairly static at approximately 20 per cent of the population. The average monthly agricultural wage – which comprised both private and state farmers – rose over the period by 62 per cent to $19. Within this, incomes of *campesino* farmers rose by 42 per cent, but remained fairly static for UBPC farmers. This average wage was still low compared to those in other productive sectors, the highest paid being the mining industry at $27 a month. Given that, on the ground, farming was one of the highest paid professions, these official figures may not include income generated by sales through non-state channels nor bonuses for over-production.

These figures for the decade of the 2000s indicate that the food security situation was stabilizing but continued to run on undesirably low and faltering levels of inputs, and still had some way to go to become more secure. In 2006, the World Wildlife Fund's Living Planet Report (WWF, 2006) identified Cuba as the only country in the world to meet its sustainable development criteria. This measurement was made by comparing Cuba's rating on the United Nations' Human Development Index – a measure of human welfare (health, education, poverty and so on) – with its ecological footprint reflected in its per

–capita fossil-fuel consumption. Cuba's apparent success was due to a combination of its socialist policy and its continued petroleum deficiencies. Because of the diversity of formal and informal food sources, food import figures were difficult to ascertain, but at 2008 were estimated to be around 80 per cent of national food needs. This percentage is high and needs to be taken in the context that the country was not attempting self-sufficiency and to some extent was making pragmatic choices over what to import and what to produce domestically. The idea of organic self-sufficiency emanated from foreigners to the country. Further, it was unlikely that this percentage took into account the multiple informal and private food supply sources.

The changes in the sugarcane sector provide an indication of the length of time necessary for a collective shift in attitude or perception. Although the inputs necessary to maintain previous levels of sugarcane production had, along with the market demand, fallen drastically in the early 1990s, it had taken until the end of the decade and into the 2000s for the state to decide to pull out of sugarcane and accept that it would never regain its previous importance. Once this realization and decision had been made, plans were put in place to plough up vast tracts of sugarcane land and plant with mixed forest, pastures and horticultural crops. Sugar had been the agricultural and economic backbone of Cuba for over a century, and was part of national identity; therefore, it was not surprising that this decision had taken over a decade to make.

A new wave of reforms at the end of the 2000s

At the end of the 2000s, more reforms were coming into place, driven by the new charge of government led by Raul Castro. Discussions on this new reform process actually commenced in 2005, but the steady stream of announcements started rolling out in 2007–08 with the stepping down of Fidel Castro, who conceded the Presidency to his brother. Many of these reforms were in the agriculture and food sectors, and several were reminiscent of efforts attempted but not achieved in the previous decade. Agriculture would be further decentralized with the opening of agricultural supply shops where farmers could directly purchase inputs, farm gate prices would be further improved, and the capped level of bonuses on top of basic salaries would be raised. Prices paid for cattle, meat and other farm products more than doubled, and all farmers owed money by Acopio were paid. Local municipal offices would take into account the activities of private farmers and cooperatives. In terms of consumption, rice cookers were handed out and microwave ovens made available for purchase. People would be encouraged to participate in debate and also to watch out for illegalities. In other sectors, public services were improved, electronic goods made available, international trade was encouraged, and a process was put in place to reduce monetary imbalance. Forecasted was the likelihood of changes to the ration system, investment in the domestic food processing industry, a greater use of incentives, further price reforms, and a reduction of benefits to the informal sector. In this sense, the state was promoting a greater sense of ownership but also attempting to clamp down on the informal sector, the black market and illegal activities (IICS, 2008; Morris, 2008).

Beyond agribusiness: why Cuba isn't farming organically

Urban versus rural agriculture

Cuba's transition towards greater food security has been built on a patchwork approach to agriculture in which organic techniques have played a significant, but not exclusive, role. Throughout the Special Period, Cuba adopted several characteristics of a localized, organic farming system. These included more location-specific strategies, a relative increase in participatory extension approaches and institutional decentralization, the promotion of organic inputs, and more localized production–consumption chains. Other characteristics were, however, barely visible, such as widespread ecological literacy, localized seed systems, collective relinquishment of industrialized practices, and the application of holistic and systems principles. A reduction in access to fuel and agrochemicals alone would not necessarily lead to widespread organic production systems. Yet the experience of urban agriculture stands as an example of what Cuban society was capable of achieving. Urban agriculture took off in the early 1990s because the desperate situation at that time encouraged an open-mindedness to innovations that worked. Within a relatively short period, small-scale, intensive urban organic agriculture showed it could deliver, and concepts such as permaculture – considered alternative in other countries – were accepted to a relatively greater extent by the mainstream. In rural areas, however, change was slower, and before organic agricultural approaches could show results, inputs were re-entering the country and the open-mindedness induced by the crisis was easing off. Significantly, urban producers frequently came from non-farming backgrounds and so were able to accept alternative concepts more easily than industrially trained agricultural professionals. Into the 2000s, those working in the urban sector remained convinced that urban production held further potential still. If it was expanded into peri-urban areas, it could, for example, provide the majority of foodstuffs for urban populations on a permanent basis. As one researcher explained, 'The organic approach will last, because even now more agrochemicals are available, yet urban agriculture is still organic. Rural farmers need to be organized to achieve the same – whole families need retraining.' Nevertheless, with the country's improved food security status, there was a lessening of state support for the further expansion of urban agriculture as compared to the 1990s.

Underlying challenges to implementing widespread organic systems

Overall, the challenges encountered by attempts to implement a more widespread organic system could be categorized as follows:

1   Lack of petrol – petrol shortages limited most agricultural activities, whether attempting to farm organically or not. This was because farms were not operating semi-closed systems, but instead were dependent on external agents for seeds, manure, pest-control inputs, irrigation turbines and so on. On all but the CCS cooperatives, traditional knowledge was either under-utilized or absent, and instead farmers relied for guidance on the mobility of agricultural advisors.

2   No policy commitment – neither the state nor the people had chosen to farm organically; the situation had been forced upon them. Out of this emerged varying degrees of reluctance to change from the previous industrial approach, and this was strongly influenced by the nationwide drive to maximize production. Any production approach had to prove itself in terms of short-term yield performance. Additionally there was a degree of institutional reticence to change, given the jobs, structures and investments already built up around the industrialized model. As one ministry staff explained, 'Even in the absence of agribusiness, industrialized agriculture in Cuba supports a bureaucratic system which does not want to change.'[2] In the absence of policy support, no markets were developed for organic produce and therefore no cost incentives existed for farmers. Additionally, activities to promote organic farming, whether by the grassroots organic movement or individual pioneers, could not flourish and could even be construed as incongruent with state policy.

3   Ecological illiteracy – although scientific and experiential evidence was showing organic agriculture to be both economically and technically sound, farmers were constrained by a lack of relevant knowledge and of organic inputs. At the same time, existing knowledge included a degree of myths and misperceptions surrounding organic production, and these perceptions influenced decisions and actions on agricultural policy and practice. Without a clear consensus over the definitions and objectives of sustainability, little concerted action could be taken in terms of policy support.

4   Beyond agribusiness – the country had no private sector, no corporate interests to advertise and promote industrial approaches and/or bypass organic. Yet even without the potential for financial gain or societal power that the

agribusiness sector in other countries are commonly considered to be driven by, an industrialized mindset pervaded in Cuba all the same. The existence of this mindset could be largely attributed to the substantial vestiges of the industrial farming system, including the formal agricultural training and accrued learning and experiences over the previous decades. Yet there was more to it than this. In the face of accepted evidence of the efficacy of organic approaches and of the degradation of the natural resource base through industrial practices, and in the near-absence of agrochemicals, an element of denial could be detected. Of the myths and misperceptions surrounding organic production, certain of these were logically grounded, but others, including its yield ability, were clearly not. A common thread of fear wove its way through the misperceptions: both a fear of losing control (over, for example, pests, diseases, farmers, nature in general), and a fear of not having enough (chemical inputs, yields, fuel and so on).[3]

Degrees of learning and transition

A consequence of these challenges was the varying degrees of learning and change experienced in the process of coping and effecting a transition from the industrialized past. At one extreme, certain practices and systems underwent very little change during the Special Period, even when they were acknowledged as being ineffective or stifling progress. These included, for example, the continued attempts in the 1990s to breed and import high-input-requiring crop hybrids, and the continued publication of recommendations for high-input agrochemical dosages. These activities continued because of the belief that the Special Period was temporary and that conditions would soon return to those of the previous, more affluent 1980s. This belief was to some extent supported by the state through its attempts to motivate the population to hold on till the situation improved, and ran alongside the contrasting advice to adapt to the leaner circumstances.[4] In other cases, considerable learning did take place, and this led to a re-evaluation of ideas and even of ideology, in order to accept the low-input conditions of the period. Examples of this include the farm re-structuring, the early development of urban agriculture, the opening of the farmers' markets, the re-introduction and promotion of ox ploughs, the uptake of participatory methodologies and the switch to hardier crops. As Deere et al commented in a review of farm organizational and market changes (1994, pp195–196), 'The Cuban government has demonstrated a good deal of flexibility with respect to long-held socialist principles regarding the organization of agricultural production,' and further, 'Cuban officials have dropped their longstanding opposition to the holding of individual usufruct parcels on State farm land.' Further into the decade, the expansion of urban agriculture, the downsizing of the UBPCs, the adjusting of farmers' markets, and so on, indicated that even more fundamental and longer-term rethinking was taking place, which entailed accepting that the new conditions were there to stay, and reformulating more intrinsic beliefs and structures to accommodate this. Some institutions and individuals remained attached to the industrialization paradigm; others moved to a substitution stage, and a small vanguard to the redesign of the production system and philosophy behind it.[5]

Future directions: the need for knowledge and intent

Future directions for agriculture in Cuba were still not completely clear, not least because of the continued dichotomy of the productionist mandate versus sustainability objectives. A large minority of farmers had the intention to become more organic and many others were ambivalent about it, being open to this direction if moves were facilitated for them. The conditions of the Special Period had exposed many in the agricultural sector to alternative directions who would not otherwise have looked for them. Policy was becoming geared towards a more integrated farming approach, although terminology was ambiguous depending on whence it was emanating. If organic agriculture is defined not only by the extent of appropriate husbandry practices but also by the intent and the knowledge base (of the farmer), then both knowledge and intent require addressing in Cuba.

Nevertheless, investment in human capacity-building takes substantial time to fully manifest and the full potential of the changes and experiences of the Special Period will continue to unfold. With its high education levels and research capacity, an absence of conflicting interests from an agribusiness sector, protection of local agricultural markets, well-endowed (if unevenly spread) natural resources, strong communication channels, a wide spread of extension agents, smaller farm sizes and more human-scale management, and an imbued socialist culture, Cuba has good conditions for mainstreaming organic agriculture. Other industrialized countries with a less-developed infrastructure, less-collective cohesion and more entrenched corporate interests would seem to face greater challenges in achieving such a transition.

Perhaps of most interest for the international furtherance of localized, organic production – and change in the face of peak oil – is the emerging challenge of the industrialized mindset as being possibly more significant than the existence of corporate agribusiness interests. In Cuba, the latter was virtually absent, but the former was strong. On the one hand, this mindset may be influenced or enlightened through increases in ecological literacy, not only for professionals working in agriculture but also those in economics. This would enable consensus on an informed, sustainable vision and would provide the principles for implementing such a vision. Increasing ecological literacy may also allay some fears, such as over the ability of organic production to provide sufficient food and over the order in apparent chaos.

Implications of the Cuban experience for global agriculture and food security

Global consensus exists over the need for widespread change in order to deal with peak oil as well as with climate change. It also exists over the need for widespread change in the food system in order to achieve food security, and in the farming system to become more sustainable. Yet although these issues have been on the public radar for at least 40 years, the desired goals and pathways to reach them are unclear, and relatively little has actually been achieved. Cuba is quite unique in its mode of centralized governance, and some might argue that because of this it is difficult to extrapolate from these experiences. Yet in almost every other part of the world, decisions over resources connected to agriculture and the food supply chain are highly centralized amongst a few corporations. The extent of real, conscious choice available to both consumers and producers may be very similar. These apparently different ideologies could in fact be stemming from the same paradigm, and Finn (1998) suggests that centralization is a practice promoted by old socialism as well as by competitive market-driven advocates, albeit that one is state-owned and the other private.

One feature of industrial farming and food systems is the increasing levels of mechanization and homogenization. These systems, with their long food-supply chains, play a large role in current patterns of fossil fuel consumption. By contrast, Cuba has to some extent been moving in the opposite direction, towards more decentralized, human-scale and bioregional production and consumption systems, with greater levels of autonomy, diversity and complexity. As and when the predicted global fuel supply crisis fully materializes, Cuba's example provides lessons as to how it might be addressed. As Snyder, a US citizen reporting back from Cuba, stated, 'Few if any advocates for sustainable agriculture in our own country would wish to swap our government or economic circumstances with those found in Cuba. But it sure doesn't hurt to see an example of how we might utilize the principles of sustainability in the United States to avoid our own Special Period in the future' (Snyder, 2003).

Cuba's achievement in moving from a highly vulnerable situation to one heading towards stability also stands in comparison with the experiences of many low- and middle-income countries struggling with long-term food insecurity. In particular, Cuba's example indicates that the Millennium Development Goal of halving the number of food-insecure people by 2015 is not an overly ambitious target, but is one that can be achieved by a firm political commitment to prioritize basic food rights and a semi-regulatory market approach. Non-socialist countries may not be immediately sympathetic with the measures that Cuba had taken to ensure equity, such as the use of rations or of prioritizing domestic markets, yet Cuba's experience has shown these measures to be viable and an arguably necessary means of assuring access to food for all during periods of vulnerability. This, in the face of fossil fuel deficits, is perhaps the biggest lesson that Cuba's experience of the 1990s has for the rest of the world.

Notes

1   According to Haverkort et al (1991), problematic impacts of paternalism include blinding people to the need for solving their own problems, accustomizing them to expect give-aways, and destroying the possibility of multiplier effects.
2   Supporting this observation, MacRae et al (1990) suggest that in the absence of explicit political objectives to promote organic agriculture, there is a form of institutional inertia whereby decision-makers tend to postpone action until there is either overwhelming scientific data to support it or overwhelming negative effects of the prolonged inaction.
3   Hill (1991) links the development of sustainable agriculture with psychological prerequisites, arguing that distressed human states, such as a fear of losing control or of not having enough, result in 'the stubborn adherence to production systems that are progressively proving to be unsustainable'.
4   Röling (2005, p100) identifies this phenomenon as a lack of 'The ability to note discrepancy, to adapt the reality world, to feedback.' Based on the constructivist perspective of reality, he explains: 'We can build a cosy coherent reality world, in which our values, theories, perceptions and actions are mutually consistent. But this reality world can become divorced from its domain of existence; for example, it can fail to correspond to ecological imperatives.'
5   A simple pressure exerted from the environment may induce a simple change in action. However, if the problem is more dramatic, then deeper-held theories and beliefs may need to be reflected upon and revised. This more fundamental change has been termed double loop learning, as a single cycle of learning and action did not effect the change required to deal with the situation (Argyris and Schön, 1996). Even more extreme change occurs through triple loop learning, whereby essential principles upon which beliefs and institutions are based come under discussion, and this occurs over a longer time period. Triple loop learning may entail an ideological change to a more adaptive management approach (Holling, 1995).

References

Antonelli, G. (1990) 'Soviet agriculture in the current transitional phase', *Questione-Agraria*, no 37, pp51–73
Argyris, C. and Schön, D. A. (1996) *Organisational Learning (II), Theory, Methods and Practice*, Reading MA, Addison-Wesley
Csaki, C. (1998) 'Agricultural research in transforming Central and Eastern Europe', *European Review of Agricultural Economics*, vol 25, no 3, pp289–306
Csizinszky, A. A. (2003) 'Developing international collaborations in Central and Eastern Europe', *Hortscience*, vol 38, no 5 (August), p656

Deere, C. D., Pérez, N. and Gonzales, E. (1994) 'The view from below: Cuban agriculture in the "Special Period in Peacetime"', *The Journal of Peasant Studies*, vol 21, no 2 (January), pp194–234

FAO (2004) *Food and Agriculture Indicators*, Cuba. www.fao.org/countryprofiles

Finn, P. (1998) 'Polish Farms Face Being Ploughed Under', *Guardian Weekly*, 1 November, p17

Granma (2007) 'Electricity production in Cuba exceeds maximum demand', *Granma International*, 4 October 2007

Haverkort, B., van der Kamp, J. and Waters-Bayer, A. (1991) *Joining Farmers' Experiments. Experiences in Participatory Technology Development*, London, Intermediate Technology Publications

Hill, S. B. (1991) 'Ecological and psychological prerequisites for the establishment of sustainable agriculture prairie communities', in Martin, J. (ed) *Alternative Futures for Prairie Agriculture Communities*, Calgary, University of Calgary

Holling, C. S. (1995) 'What barriers? What bridges?' in Gunderson, L. H., Holling, C. S. and Light, S. S. (eds) *Barriers and Bridges to the Renewal of Ecosystems and Institutions*, New York, Columbia University Press

IICS (2008) *Cuba Reforms Summary 2008*. Press release, International Institute for Cuban Studies, London Metropolitan University

Little, R. E. (1998) 'Public health in central and eastern Europe and the role of environmental pollution', *Annual Review of Public Health*, no 19, pp153–172

MacRae, R. J., Hill, S. B., Henning, J. and Bentley, A. J. (1990) 'Policies, programs and regulations to support the transition to sustainable agriculture in Canada', *American Journal of Alternative Agriculture*, vol 5, no 2, pp76–92

Molnar, I. (2003) 'Cropping systems in Eastern Europe: past, present and future', *Journal of Crop Production*, vol 9, no 1/2, pp623–647

Morris, E. (2008) 'After Fidel: prospects for economic change', *After Fidel: The Prospects for Change in Cuba and US/Cuba Relations*, London, Canning House

ONE (2006) *Anuario Estadístico de Cuba 2006*. Edicion 2007, La Habana, Oficina Nacional de Estadisticas

Patsiorkovski, V. V. and O'Brien, D. J. (1997) 'Material changes, subjective quality of life, and symptoms of stress in three Russian villages', *Journal of the Community Development Society*, vol 28, no 2, pp170–185

Röling, N. (2005) 'The human and social dimensions of pest management for agricultural sustainability', in Pretty, J. N. (ed) *The Pesticide Detox: Toward a More Sustainable Agriculture*, London, Earthscan Publications, pp97–115

Rosset, P. M. (1996) 'Cuba: alternative agriculture during crisis', in Thrupp, L. A. (ed) *New Partnerships for Sustainable Agriculture*, World Resources Institute, Washington DC, pp64–74

Sinclair, M. and Thompson, M. (2001) *Cuba Going Against the Grain: Agricultural Crisis and Transformation*, Oxfam America. http://www.oxfamamerica.org/newsandpublications/publications/research_reports/art1164.html accessed May 2002

Snyder, B. (2003) *Cuba: a Clue to Our Future?* Oakland CA, Food First/Institute for Food and Development Policy. www.foodfirst.org/media/news/2003/cubacluetofuture.html

Turnock, D. (1996) 'Agriculture in Eastern Europe: communism, the transition and the future', *Geojournal*, vol 38, no 2, pp137–149

Wiskind, A. (2007) 'Cuba: sustainability pioneer?', *World Watch Magazine*, 1 July 2007. http://goliath.ecnext.com/coms2/gi_0199–6707014/Cuba-sustainability-pioneer-FROM-READERS.html#abstract

WWF (2006) *Living Planet Report 2006*, Switzerland, WWF International. http://assets.panda.org/downloads/living_planet_report.pdf

Appendix A

Extent of Use of Organic Techniques by Farmers Surveyed

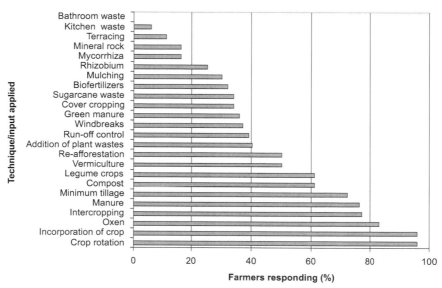

Figure A.1 *Degree of use of organic techniques and inputs to improve soil fertility*
Source: Wright, 2005

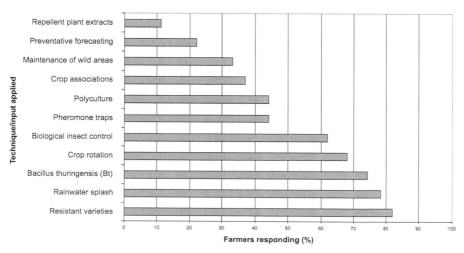

Figure A.2 *Degree of use of organic techniques and inputs to control pests and diseases*

Source: Wright, 2005

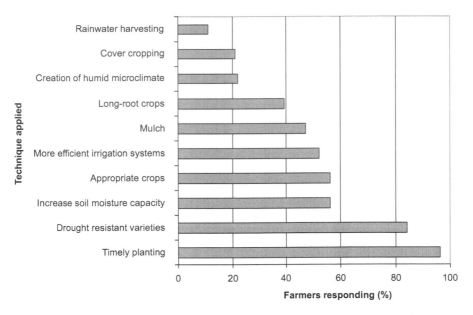

Figure A.3 *Degree of use of alternative water management and conservation techniques*

Source: Wright, 2005

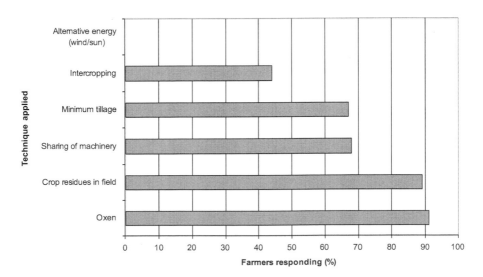

Figure A.4 *Degree of use of energy-efficient labour techniques*

Source: Wright, 2005

Reference

Wright, J. (2005) *Falta Petroleo. Perspectives on the Emergence of a More Ecological Farming and Food System in Post-Crisis Cuba.* Doctoral thesis, Wageningen University

Appendix B

Agriculture and Food Related Institutions in the Early 1990s

The organization of agricultural institutions

Administratively, the country was organized into 14 provinces containing 169 municipalities and many more People's Councils (*consejos populares*) at the lowest administrative level. Given Cuba's planned economy and infrastructure, the agricultural support system had a clear structure within this. Two Ministries were concerned with agricultural production: the Ministry of Agriculture (MINAG) and the Ministry of Sugar (MINAZ). Although they were institutionally separate, some non-state farms produced for both. A basic organogram of the ministerial organizational structure is shown in Figure B.1.

The Ministry of Agriculture (MINAG) and its entities

The Ministry of Agriculture contained 24 departments and 19 national-level research and support institutions that comprised the National Institute of Plant Protection (Sanidad Vegetal), the Soils Institute (Instituto de Suelos), and 17 largely commodity-focused agricultural research centres. For production support and marketing, MINAG operated a nationwide network of institutional entities: Agricultural Enterprises, the Cooperative Campesino Sector (Sector Cooperativa Campesino, SCC) and Acopio (described later), as well as running 222 state farms (MINAG, 1999). The Credit and Commerce Bank (BANDEC) provided credit to the state farms and cooperatives.

State Agricultural Enterprises (Emprezas)
State production was coordinated by various provincial Agricultural Enterprises; each one specialized in a specific production component or commodity, such as mixed crops, livestock or seeds. There were 487 throughout the country (MINAG, 1999).

Campesino Cooperative Sector (SCC)
This organization worked closely with non-state producers, to sell seeds, work clothes and small tools, and other inputs.

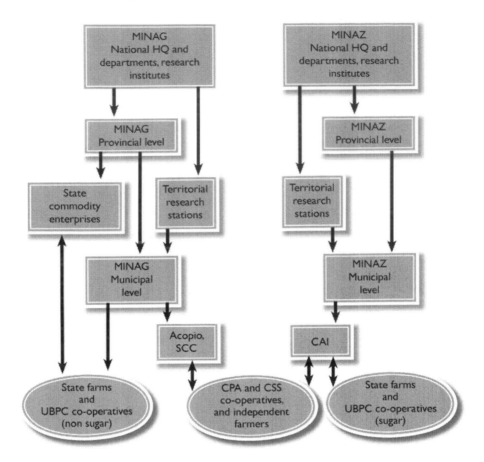

Figure B.1 *Organogram of the institutional structure of the Ministry of Agriculture and the Ministry of Sugar*
(Note: arrows indicate flows of policy direction, technology, information or market support)

Source: Miedema and Trinks, 1998

Acopio[1]

The state food collection and distribution enterprise, Acopio, was responsible for making production/delivery plans with farmers, and for the collection, transport and distribution/marketing of produce for the ration and to other outlets. It worked with both state and non-state producers. Almost 100 per cent of state production and about 50–80 per cent of non-state production was moved through Acopio (Oliveros Blet et al, 1998).

National Institute of Plant Protection (Sanidad Vegetal)

This family of institutes was dedicated to plant health, embracing pest and disease control; policy, legislation and control of compliance on phytosanitary issues; assistance and control on production of biological pest control products; research on plant protection; and information diffusion to government and producers. It had 257 staff, of whom 84 per cent were researchers, and its research operations were headed by the National Research Institute of Plant Protection (INISAV). Recommended pesticide and herbicide dosages for different crops were decided at national level and disseminated in the form of provincial guidelines.

Territorial Stations of Plant Protection (ETPP)

These were initially formed to provide an early warning system of pest and disease outbreak through analysis of meteorological conditions. They worked with on-farm counterparts or *activistas* to check on pests and diseases, provide advice and assistance, and ensure that phytosanitary laws were obeyed. Farmers were obliged to adhere to these laws and had to inform Sanidad Vegetal of any pest or disease outbreaks (from Decree Law No. 153, 31 August 1994). Phytosanitary specialists from the ETPPs had sole entitlement for 'prescribing' chemical inputs, and with such authorization farmers then accessed these from the agrochemical warehouse or Enterprise.

CREEs and Bio-preparation Plants

Officially within Sanidad Vegetal, the Centres for the Reproduction of Entomophages and Entomopathogens (CREE) produced biological control products, mainly *Beauvaria*, *Bacillus*, *Metarhizium*, *Trichoderma*, *Trichogramma* and *Verticillium*. Established in 1992, there were 192 CREEs by the following year, and by 1998 there were 222 (Miedema and Trinks, 1998). The CREEs were situated on farms, as stand-alone establishments, or in higher education establishments. Depending on location they might be staffed by technicians and/or high school graduates. Cuba also had three larger factories dedicated to the production of the bacterium *Bacillus thuringiensis*, the most widely used biological control. Product prices were lower than those of chemical pest- and disease-control products, and were sold in Cuban pesos.

Soils Institute (Instituto de Suelos)

Originally the Research Institute for Soils and Agrochemicals, this was established in 1977 to undertake research and service provision, and carry out state functions. As well as provincial offices, the Institute had three regional experimental stations. Organic soil fertility inputs included bacteria such as *Azospirillum*, *Azotobacter* and *Rhizobium*, and fungi such as *Mycorrhiza*. Their uses were recommended for a wide range of crops including sugarcane, cassava, onion, citrus and maize (Miedema and Trinks, 1998). Also falling under the Soils Institute were Basic Units of Worm Culture (UBL).

Other ministries relevant to agriculture

Ministry of Sugar (MINAZ)

The existence of a Ministry of Sugar highlights the importance of sugar in the Cuban economy. It was responsible for an area of over 1.5 million ha of productive land, dealing with both state and non-state producers. It operated Agro-Industrial Centres (CAI, Centro Agro-Industrial), which fulfilled similar functions to MINAG's Acopio and the Agricultural Enterprises, and it also ran sugarcane processing factories. As an important national export product, sugar received state priority for petrol, fertilizers and herbicides (Miedema and Trinks, 1998).

Ministry of Higher Education (MES)

Several important agricultural research institutes fell under the Ministry of Higher Education. These included the National Agricultural University of Havana, UNAH (previously known as ISCAH), which housed the Centre for Studies on Sustainable Agriculture (CEAS), and the National Institute of Agricultural Sciences (INCA) (which hosted the author's research project). INCA and UNAH jointly published, from 1967 onwards, the *Revista Cubana de Ciencia Agrícola*.

Ministry of Science, Technology and the Environment (CITMA)

This Ministry emerged in 1994 out of the Cuban National Academy of Sciences. Amongst other roles, it was responsible for environmental education and for environmental laws, some of which affected the agricultural sector.

Major non-state agricultural institutions

National Association of Small Farmers (ANAP)

Formed in 1961, ANAP's main initial task was 'To inform farmers about ideological goals of the Revolution, and to prepare them against an imperialist attack and counter-revolutionaries.' Though officially independent from the state, it followed the line of the Cuban Communist Party (PCC) and was seen as the political organization of non-state producer cooperatives – the CPAs, CCSs and individual farmer members. It was well represented on Cuba's National Assembly, with 14 members among the 490 representatives. ANAP's main tasks were to oversee production plans, represent the social and economic interests of the farmers and provide practical assistance.

Note

1   Derived from 'acopiar', meaning 'to store or collect'.

References

Miedema, J. and Trinks, M. (1998) *Cuba and Ecological Agriculture. An Analysis of the Conversion of the Cuban Agriculture towards a more Sustainable Agriculture.* Unpublished MSc thesis, Wageningen Agricultural University

MINAG (1999) *Datos básicos*, Havana, Ministerio de la Agricultura

Oliveros Blet, A., Herrera Sorzano, A. and Montiel Rodríguez, S. (1998) *El Abasto Alimentario en Cuba y Sus Mecanismos de Funcionamiento.* Unpublished manuscript, Faculty of Geography, University of Havana

Index

20 years of publishing
for a sustainable future

The Transformation of Agri-Food Systems

Globalization, Supply Chains and Smallholder Farmers

Ellen B. McCullough, Prabhu L. Pingali and Kostas G. Stamoulis

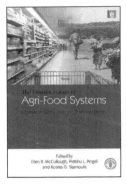

'There should be a good market for this book. The topic is very timely and a major theme of the new World Development Report 2008. The editors and contributors are world class.' *Derek Byerlee, World Bank*

'This is a topic of wide interest and high policy importance. The depth of coverage and excellent synthesis should ensure that the book will have a substantial market in high-level undergraduate and graduate courses in agricultural development. It will have a solid readership among development economists and policy makers as well.' *Mark Rosegrant, International Food Policy Research Institute*

The driving forces of income growth, demographic shifts, globalization and technical change have led to a reorganization of food systems from farm to plate. The characteristics of supply chains – particularly the role of supermarkets – linking farmers have changed, from consumption and retail to wholesale, processing, procurement and production.

This has had a dramatic effect on smallholder farmers, particularly in developing countries. This book presents a comprehensive framework for assessing the impacts of changing agri-food systems on smallholder farmers, recognizing the importance of heterogeneity between developing countries as well as within them.

Published with FAO.

Ellen B. McCullough and **Prabhu L. Pingali** were, at the time of preparation of this book, economists at the Food and Agriculture Organization (FAO) of the United Nations, Rome, Italy. **Kostas G. Stamoulis** is an economist at FAO, Italy. **Prabhu Pingali** is the former president of the International Association of Agricultural Economists.

Paperback £39.95 • 416 pages • 978-1-84407-569-0 • August 2008

For more details and a full listing of Earthscan titles visit:

www.earthscan.co.uk

Printed in the USA/Agawam, MA
April 16, 2012

565367.103